THE NUTS AND BOLTS OF ORGANIC CHEMISTRY:
A Student's Guide to Success

JOEL KARTY

Elon University

PEARSON

Benjamin Cummings

San Francisco • Boston • New York
Cape Town • Hong Kong • London • Madrid • Mexico City
Montreal • Munich • Paris • Singapore • Sydney • Tokyo • Toronto

Acquisitions Editor: Jim Smith
Project Manager: Cinnamon Hearst
Managing Editor, Production: Erin Gregg
Production Supervisor: Vivian McDougal
Production Management and Art Coordination: Heather Meledin, Progressive Publishing Alternatives
Illustrations: Progressive Publishing Alternatives
Marketing Manager: Scott Dustan
Compositor: Progressive Information Technologies
Text and Cover Design: Progressive Publishing Alternatives
Manufacturing Buyer: Michael Early
Cover Printer: Phoenix Color
Printer: RR Donnelley

ISBN 0-8053-3117-4

Library of Congress Cataloging-in-Publication Data

Karty, Joel.
  The nuts and bolts of organic chemistry: a student's guide to
success/Joel Karty.
    p. cm.
  ISBN 0-8053-3117-4
  1. Chemistry, Organic—Textbooks. I. Title.
  QD253.3.K37 2006
  547—dc22

                        2005005083

PEARSON

Benjamin
Cummings

To Valerie, for your love, patience and support.

To Joshua and Jacob, may you always ask questions.

CONTENTS

CHAPTER 8   *$S_N1/S_N2/E1/E2$ REACTIONS: THE WHOLE STORY*   169

CHAPTER 9   *CONCLUDING REMARKS: WHAT NOW?*   193

FOREWORD

TO THE STUDENT

Ask any non-science student if they have heard of organic chemistry. Invariably they have. Chances are that they will also tell you they have heard it is "hard." Clearly, then, organic chemistry has a reputation. That reputation, in conjunction with its perceived importance for admission into medical school, as well as dental, pharmacy, and veterinary school, has spurred the emergence of numerous supplemental materials for the course: supplementary organic texts that are more explanatory than traditional textbooks; others that simply distill a 1000-page textbook down to a 200-page paperback, often making the material even more difficult to understand; still others that are centered around providing short-cuts. The trouble with those supplements that provide shortcuts is that they only work for a small portion of organic chemistry, at best, and end up leading students astray for the rest of it.

Despite their different approaches, each type of supplement mentioned above shares at least one thing in common: they are all designed to be used DURING, and THROUGH-OUT the semester. In my experience as a professor, however, waiting to build critical thinking skills after the semester begins turns out to be costly for a large percentage of students. The primary reason is that students quickly become overwhelmed with what appears to be a vast amount of disconnected information (the good news, as I explain in Chapter 1, is that it only *appears* to be a vast amount of material, and it only *appears* to be disconnected). Approached in the right way, though, organic chemistry can make a lot of sense.

This text is unique. It is designed to be used BEFORE diving too far into a traditional text-book—often this means before the semester begins—in order to prepare you with the right mindset for learning organic chemistry. That mindset is essentially one that focuses you on a small number of straightforward, fundamental concepts, and asks you to apply them in different ways to solve the variety of problems you will face. Unfortunately, convincing you to take such an approach requires more of me than simply telling you that fundamental concepts are important—I don't expect you to trust me unconditionally. Instead, convincing you requires me showing you. Hence the remainder of the book.

In each chapter, a fundamental concept is fully developed in a manner that is rather easy to follow. Applications are then presented at the end of each chapter to exhibit the utility of the fundamental concepts, showing that first, most of organic chemistry is well described by a handful of fundamental concepts, and second, many problems in organic chemistry are much more closely related than they might initially appear. As a result, you come to understand that what otherwise seems like an overwhelming amount of information is actually quite palatable.

Of course there is a substantial amount of organic chemistry presented throughout this text. Keep in mind that it is presented, first and foremost, to display the power of learning and understanding the fundamental concepts. Yes, the organic chemistry itself is important, but you can trust that you will revisit the actual chemistry in your full year of organic chemistry. What is most important is that you take with you the recognition that fundamental concepts are crucial for success in this course, along with a good sense for how to apply those concepts to various situations.

TO THE PROFESSOR

It is not difficult to convince organic chemistry professors that the key to success in the course is understanding reaction mechanisms. After all, it is the mechanism that is the natural tie between reactions, and understanding what drives a mechanism allows one to make predictions about new reactions. The problem is convincing the students, and convincing them early. I certainly have had difficulty with this in my own classes. Despite a substantial emphasis on the mechanism in the textbook, and despite my own echoing of the importance of the mechanism, memorization remains quite attractive to students. I have therefore written this text primarily to convince students that learning and applying fundamental concepts and fundamental mechanisms is the key to success in organic chemistry. In order to accomplish that goal, I introduce fundamental concepts in a way that students can grasp, and I explicitly show the power of understanding each concept through application toward a variety of different aspects they will face throughout the course.

In organizing this text, I asked myself why it is that students so adamantly resist the mechanism. Could it perhaps be related to the fact that most textbooks are organized by functional group? Although there is nothing inherently wrong with organizing a textbook that way, it may be that we are effectively de-emphasizing the importance of the mechanism by unduly, albeit unintentionally, stressing the importance of functional groups. Without a clear and unambiguous message as to the true natural tie between reactions (i.e., mechanisms), this may, in fact, promote the use of flashcards and memorization.

Could the problem be that, with an organization by functional group, there are too many distractions in each chapter? For example, in a given functional group chapter, a student may be presented with nomenclature rules, physical properties, new concepts, new mechanisms, new reaction types, and new spectroscopic information. How is a student supposed to glean the important material, let alone digest it? How are they supposed to integrate what they learn about reactions in the new chapter with what they have learned in previous chapters?

Is the problem perhaps related to introducing reactions (including predict-the-product and synthesis problems) too early? Students seem to be naturally drawn to reactions. In fact, I see most students enter the course with a pre-conceived notion that predicting the products of reactions is the "end-all and be-all" of organic chemistry. Therefore, if we introduce new reactions simultaneously with other material that we deem important (such as the mechanisms of that reaction), will students overlook the other material, and simply focus on predicting the products of that new reaction? In other words, will they simply try to memorize those new reactions?

This text is designed to combat the above problems in a variety of ways. Instead of being organized by functional group, the chapters are organized by fundamental concept. In doing so, I believe that the chapters essentially tell a coherent story of reactivity and reactions in organic chemistry. Material that does not fit nicely within that story is removed from the chapter, and is either introduced as an end-of-chapter application or is not included at all. For example, physical properties are introduced as applications of intermolecular interactions (Chapter 7), whereas nomenclature and optical activity are omitted entirely; they are left for the traditional textbook. This should dramatically cut down on the distractions that students face, and at the same time it should emphasize the importance of learning and understanding each fundamental concept. Furthermore, reactions are de-emphasized by postponing them (Chapter 8) until after students have a solid foundation of mechanisms.

Mechanisms are not introduced until fairly late in this text—Chapter 6. Mechanisms are central to organic reactions, but I feel that they should not be introduced until students have a solid understanding of the fundamental concepts. The reason is that in order to understand a reaction mechanism fully (that is, *why* each step occurs as it does, and not some other step the student might imagine), a student must first understand those fundamental concepts and how they are applied. Otherwise, "learning" mechanisms becomes rote memorization, and a student gains no predictive capabilities whatsoever.

Given that the central idea of this text is shaping a student's approach toward organic chemistry, the earlier the text is introduced the better. The specific way in which the text is used, however, depends upon whether or not you envision incorporating the organizational ideas espoused by the text into your lectures. Should you make this text part of your course, you might wait until after completing it to dive into the textbook. This, in fact, is what I do in my own course and it has proven quite successful (if interested in discussing this further, feel free to email me at jkarty@elon.edu). Should you, instead, begin coverage of the textbook on or shortly after the first day of lecture, this text might best serve as recommended or required supplemental reading to be completed by your students in the first few weeks.

A novel idea is to *recommend this text to your students before the semester begins*. Some students planning to enroll in your class might approach you at the end of their general chemistry course, asking if there is anything they can do in preparation for organic chemistry (I routinely have students that do this). It would certainly be easy to direct those students toward this book, as summer reading—many students could benefit greatly from reading this text before the first day of class.

Why Do Most Students Struggle with Organic Chemistry?

1.1 INTRODUCTION

The complete and utter impossibility of organic chemistry is a complete and utter misconception. So why is it, then, that organic chemistry is *the* class? It is *the* weeding-out class; it is *the* most difficult course required for premedical students and others preparing for the allied health profession; it is *the* class that is most heavily looked at by medical school admissions committees. From my experiences, both as a student and as a professor, I find that the answer is *not* that organic chemistry is intrinsically so difficult, but rather, organic chemistry is simply *different* from any class you have ever taken. Many students struggle because they fail to see that it is such a different course by its very nature, and they fail to see exactly how it is different. Therefore, the goals of this chapter are: (1) to provide insight into how organic chemistry differs from other classes you have taken and may have excelled in; (2) to convince you that the seemingly obvious way to approach organic chemistry will, in actuality, be your downfall; and (3) to provide you with the right mind-set for organic chemistry and to explain, to some extent, the right strategy for the course.

1.2 HOW ORGANIC CHEMISTRY IS DIFFERENT

Organic chemistry is different because it probably demands more analysis, critical thinking, and reasoning than any class you have previously taken. Although most students don't realize it, organic chemistry is much more a course in problem solving than anything else. In fact, it is for this reason that a friend of mine, who is a biology professor, believes that organic chemistry is the best, and the single most important course, a biology major can take. Not because of the specific applications it has in biology, but rather because organic chemistry exposes students to new ways of thinking. It is the ability to think critically that has application to biology—and elsewhere.

Medical schools take much the same view. The grade that you earn in organic chemistry will likely be singled out by any medical school you apply to, perhaps making this grade the single most important one on your transcript. Many students believe this is because medical schools use that grade to see how well you "perform" under pressure and stress. While there may be some truth to this, medical schools are much more interested in knowing how well you can think and how you apply what you already know to new situations. Of the course grades that typically appear on transcripts in their applicant pools, they believe that the organic chemistry grade provides them the most insight.[1] I believe they are correct.

The way the MCAT is written reflects the importance of critical thinking and problem solving in organic chemistry. Just turn to the biological sciences portion of an MCAT exam book and look at some of the organic questions—especially the ones pertaining to a passage. These questions may ask you why it is, based on the results of some experiment described in the passage, that Scientist A believes that the reaction occurs by Scheme I instead of Scheme II. Or you may find a question asking what the product of Reaction IV would be if one of the reagents was replaced with another similar reagent. These questions are not simply asking for regurgitation of information. In fact, these questions often cover material you are not expected to have seen before. They are asking you to *extend* the fundamental concepts you learn in organic chemistry to new situations and new problems. This is no accident; the American Association of Medical Colleges—the organization that composes and publishes the MCAT—specifically does not want the next generation of doctors to be good at only regurgitation. They maintain that the amount of knowledge out there is simply too great to memorize, and doctors are routinely called upon to extend the ideas that they do know to new problems and new situations.[2]

1.3 THE WRONG STRATEGY

Before we discuss "the right strategy" for organic chemistry, we should first discuss the wrong strategy, in hopes of dispelling a common but grave misconception. This misconception is that organic chemistry requires an inordinate amount of memorization. The reality is that it most certainly does *not*—and the more you rely on memorization in this course, the worse you will do!

---

[1] *Brenda Armstrong, director of admissions, Duke University School of Medicine.*

[2] *Lois Colburn, assistant vice president, AAMC Division of Community and Minority Programs*

Despite reality, the vast majority of organic chemistry students each year enter the course with the idea that memorization is the key—so much so that they quickly resort to flash cards, either homemade or store bought. Now, that's not to say that flash cards are necessarily evil or that they are not useful. Instead, what I have observed, time and again, is that flash cards are misused. For a given reaction, the tendency is to memorize the reactants, the products, the necessary reaction conditions, and a few other things. As a result, focus is removed from understanding *why* it is that that reaction forms the products it does, as well as *why* it is that changing the reaction conditions only slightly may yield an entirely different product! Understanding why such things happen is far more useful, for the following reasons:

1 Memorizing is "easy in, easy out." If you simply memorize a reaction without understanding why those particular products are formed, or why those reagents are necessary to carry out that reaction, or what drives the reaction, then you will quickly forget whatever you memorized. There is no foundation. Consequently, when it comes to your final exam for Organic I, which will likely be cumulative, you will have to memorize a second time. Several Organic II courses use final exams that are cumulative over the entire year, so you may have to memorize a third time. Finally, it could be a year or longer between the end of your organic chemistry course and the MCAT (or other standardized test) exam date. You will have to memorize yet again. On the other hand, if you understand a reaction to greater depth than just being able to regurgitate reactants and products, your retention of that reaction (and everything about it) will be much longer. Instead of having to memorize several times, you must learn only once.

2 In organic chemistry, there are a handful of fundamental concepts that can explain a wide variety of chemical reactions and phenomena. These reactions and phenomena are tightly connected, even though they may appear to be quite disparate. Therefore, if you rely heavily on memorization, then each of the reactions and phenomena that you encounter throughout a year of organic chemistry (all told, perhaps hundreds) will remain independent, disconnected pieces of information. That represents *a lot* of effort, for little gained.

3 The MCAT and other standardized exams seem to expect that the majority of students rely on memorization. As I alluded to earlier, in order to make distinctions among students in a competitive and talented pool, their questions are often designed to favor those who understand the few basic concepts and who can extend that knowledge. In other words, such questions are probably designed to distinguish between those who relied on memorization and those who did not.

Memorizing is attractive. But if you want to succeed in this course, you must keep it to an absolute minimum. As a student, I found organic chemistry to be one of the most straightforward courses in college, and I never wrote or made use of any flash cards. As a professor, I observe that my top students, year in and year out, refrain from memorizing and using flash cards. Conversely, those students who tell me that they used flash cards in studying are the ones who found organic chemistry indescribably frustrating.

One of my favorite stories as a professor involves such a notion. One year, a student taking my Organic Chemistry I course earned a C on the first hour exam, an F on the second, and a D on the third. Immediately after the third exam, she came to my office, quite frustrated and overwhelmed, proclaiming that she was going to drop the course; she couldn't stand any more frustration, even though we were only 3 weeks away from finishing the semester—one more exam plus the final exam. It took me only a few minutes to convince her not to drop and to finish out the semester. To help solve the problem, I asked her how she studied. Sure enough, flash cards and memorization were among the first things she mentioned. I asked her: "Why the flash cards and memorization, even though I warned you, and even though I showed you all along how to use the right strategy?" I was quite impressed with her answer: "How do you expect us to undo what we've been trained to do for 18 years?" Although this would be a good end to the story, there's one more piece to it. I spent the next hour reviewing with her some of the basic strategies I had stressed all semester. After that meeting, I didn't see her again in my office. She earned a B+ on that fourth hour exam, and shortly thereafter, a B on the cumulative final exam. The following semester, in Organic II, this same student earned a 96 percent on one of the exams—the highest grade in the class and an incredible turnaround.

1.4 THE RIGHT STRATEGY

I've spoken quite a bit about the pitfalls that accompany memorization. It is clear to me, and to many whom I have taught, that memorizing your way through organic chemistry is the wrong strategy. So what is "the right strategy?" The right strategy is to focus primarily on learning and understanding fundamental concepts that underlie organic chemistry and to gain competency in applying them to each chemical reaction and phenomenon you encounter. You might guess that this would yield results that are no better than memorization and would be more time-consuming. On the contrary, it will be easier, less frustrating, and a time saver in the long run. Instead of memorizing hundreds of reactions—their reactants, products, reaction conditions, and so on—you need only learn and understand the small number of fundamental concepts and be able to apply them to a large number of situations.

Let me elaborate on this using an example from geometry (I choose geometry, in part because it tends to elicit many of the same feelings that organic chemistry does). Suppose that in class, your professor covered interior angles of regular polygons (all sides and angles identical). Later, on the exam, you are faced with the following question: "What is the interior angle in a nine-sided regular polygon? There are three ways you might imagine having studied for a question like this. First, you could have memorized the interior angles of as many regular polygons as possible—60° for a triangle, 90° for a square, 108° for a pentagon, 120° for a hexagon, etc. This is a mistake because (1) there is a limit to the number of angles that you can memorize, and you may not have been able to memorize the angle of the nine-sided polygon; and (2) even if you did memorize it before the exam, you run the risk of drawing a blank come exam time.

A second strategy is to memorize the equation that yields the interior angle of an *n*-sided regular polygon, as shown in Equation 1-1:

$$\theta = 180° - (360°/n) \tag{1-1}$$

where θ is the interior angle. The advantage is that memorizing this equation cuts down significantly on the amount of material that you must memorize. However, the problem with this strategy is that it is still memorization. Therefore, you still run the risk of drawing a blank on the exam, particularly because there are numerous other equations that you had to memorize while studying. Or, perhaps you remember the equation's general form but you don't remember the details, such as whether 180 is outside the parentheses and 360 is inside, or vice versa. Both scenarios result from not having any context for the equation—you don't know why it works, but it provides the correct answer.

The best strategy to ensure success on a problem like this is to learn and understand simple fundamental concepts that are easy to grasp and which allow you to derive the answer quickly and in a way that makes sense. Because the concepts and the derivation make sense, the risks involved with memorization are eliminated. Furthermore, your retention of the material will be lengthened tremendously.

For our problem, there are two fundamental concepts to apply: (1) there are 360° in a circle, and (2) there are 180° in a line. Imagine, then, driving a car along the perimeter of a regular polygon with *n* sides (Figure 1-1). You end up driving in a straight line until you encounter a vertex of the polygon, at which point you make your first turn. The angle by which you change direction will be an *exterior angle* of the polygon. After completing the turn, you again drive in a straight line until you encounter the second vertex and make the second turn. Eventually, you end up at your original position after having made *n* turns, at which point you will have turned a total of 360° (the total number of degrees in a circle). Therefore, each turn you made must have been (360°/n)—that is, each exterior angle must be (360°/n).

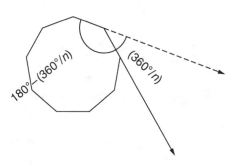

Figure 1-1 Derivation of the interior angle of an *n*-sided regular polygon. The exterior and interior angle sharing the same vertex must sum to 180°.

Finally, to calculate the interior angle, we notice that the exterior and interior angles sharing the same vertex define a line and therefore must sum to 180°. This is shown in Equation 1-2, where $\theta_{external}$ and $\theta_{internal}$ are the exterior and interior angles, respectively. Substituting in $(360°/n)$ for the exterior angle, and rearranging the equation, the result is Equation 1-1.

$$\theta_{exterior} + \theta_{interior} = 180° \tag{1-2}$$

$$\theta_{interior} = 180° - \theta_{exterior} \tag{1-3}$$

$$\theta_{interior} = 180° - (360°/n) \tag{1-4}$$

The approach we just used in the geometry problem is one that is widely applicable throughout organic chemistry. Just as we employed fundamental, easy-to-grasp concepts to derive the previous answer, there are a handful of fundamental concepts in organic chemistry that we will learn, understand, and apply toward a variety of scenarios.

1.5 ORGANIZATION AND GOALS OF THE BOOK

The material in this book is essentially the same as that in my lectures for a 6-week organic chemistry prep course that I teach in the summers at the Duke University School of Medicine. Students in that class come from around 40 different colleges and universities from around the country. The following fall, they return to their respective institutions to take their full year of organic chemistry. From the organization and focus of my course, students invariably feel well prepared for, and confident about, the organic chemistry challenges they will face.

The remaining chapters of this book are organized quite differently from the traditional textbook of 1,000 pages or more. The traditional textbook is typically divided into chapters that focus primarily on specific types of compounds or specific reactions. The chapters in this book, however, are organized by specific fundamental concepts. The first several parts of each chapter introduce the concept(s); toward the end of each chapter, we examine specific examples of the application of that concept toward a variety of organic reactions and phenomena. Several of these applications will be among the most difficult you will encounter in organic chemistry.

Organizing this book in such a fashion will hopefully (1) demonstrate the importance and the power of understanding each concept instead of memorizing, (2) make connections between seemingly different reactions much more transparent than studying organic chemistry one reaction type or molecule at a time, and (3) eliminate much of your fear about your upcoming organic chemistry course.

Of the fundamental concepts we examine, probably the most important one is the *reaction mechanism*, discussed in Chapter 6. Reaction mechanisms are at the center of organic chemistry; a reaction mechanism provides a detailed, step-by-step description of what happens behind the scenes in an overall reaction. Most importantly, reaction mechanisms allow us to understand *why* reactions occur as they do. They are analogous to our exercise in deriving Equation 1-1, where we learned *why* the equation takes the form it does.

Despite the importance of reaction mechanisms, they are introduced somewhat late in this book, because understanding reaction mechanisms requires that you first be comfortable with a handful of fundamental concepts. Those concepts are introduced in Chapters 2 through 5.

In Chapter 7, we examine intermolecular interactions. As we will see, intermolecular interactions provide insight into physical properties like boiling points, melting points and solubilities. Perhaps more importantly, intermolecular interactions provide insight into effects that solvents play in chemical reactions. Understanding those solvent effects allows us to augment what we learn in some of the previous chapters.

Chapter 8 is different from the other chapters. Instead of focusing on specific concepts, it focuses on a specific set of reactions. This chapter is included because these reactions are typically the first truly new organic reactions encountered in a full year of organic chemistry. Typically, they are introduced midway through the first semester. The good news is that individually, these reactions are relatively straightforward to deal with. The not-so-good news is that there are subtle differences between them, making it difficult for many students to determine which reaction is favored under which set of conditions. As a result, these reactions alone often cause students' grades to plummet. Chapter 8 demonstrates how several of the concepts from previous chapters can be brought together to help make quite a bit of sense out of this set of reactions. In so doing, when you encounter them in your organic chemistry course, you should feel quite confident and competent.

In order to benefit you as much as possible, I've tried to make this book accessible. I include novel analogies, and I personify molecular species when talking about the energetics of chemical processes, using language like "happy," "unhappy," "wants to," "likes," and "doesn't like." In taking this approach, anyone who has completed a year of general chemistry should be comfortable reading this book. In fact, this book has already been implemented in my organic prep course I previously mentioned. Most students in that class have completed one full year of general chemistry, a significant percentage have had only the first semester, and several of them have not taken any general chemistry in college. The feedback was outstanding—they felt my book was quite readable, and many claimed that they *finally* understood some of the general chemistry concepts they didn't understand the first time around.

The intended audience for this book is anyone who has had a full year of general chemistry in college. However, because the level of rigorousness of a general chemistry class is different at each college and university, Chapters 2 and 3 are designed to get everyone up to speed. You might be tempted to skip those chapters or to lightly skim over them. I recommend not doing so. Read through them carefully. Much of the material will be presented differently from what you have seen before, which will provide additional insight and a deeper understanding. Furthermore, the discussions in those chapters are more focused on the types of problems that you encounter in organic chemistry. Finally, and perhaps most importantly, we spend time on a number of trouble spots that tend to plague the majority of students throughout organic chemistry.

1.6 YOUR JOB

There are many instances throughout this book in which I work through sample problems; there are other instances in which I will simply make a statement, having previously solved a similar problem. If you read through this book without working through the problems, you will not learn much and you will be unable to do the problems when it matters. The same is true if you take all of those statements on faith. Your job, therefore, is to be active. Get a pencil and a piece of paper and work through the problems as I present the solutions (I suggest you begin with revisiting the geometry example from this chapter). Don't believe everything that I say—prove it to yourself! This ensures a deeper understanding and a longer retention of the material. Most importantly, stay focused, especially on learning and understanding over memorization.

Happy reading!

Lewis Dot Structures and the Chemical Bond

2

2.1 INTRODUCTION

Organic chemistry is predominantly concerned with the reactions that organic molecules undergo. A **chemical reaction** can be defined as the conversion of reactants to products through the *breaking and formation of chemical bonds between atoms*. It is therefore clear that the way in which atoms bond together to form molecules is central to the chemical behavior of those molecules. In this chapter, we discuss why molecules are constructed the way they are, focusing on Lewis structures and the theory of resonance. In doing so, we also delve into formal charge and oxidation state.

The end-of-chapter applications of the principles we cover in this chapter are heavily geared toward organic problems. In Section 2.7, we will learn how to track formal charge quickly and easily throughout several steps of a chemical reaction. Section 2.8 focuses on redox reactions, with specific application toward predicting reasonable products. Finally, in Section 2.9, we work through exercises in drawing resonance structures of molecules.

2.2 ATOMS AND ELECTRON CONFIGURATION

Before we begin our discussion of bonding between two atoms, we must review the behavior of electrons in single atoms. After all, the type of bonding that we will be most concerned with, covalent bonding, involves the sharing of electrons between the atoms from which those electrons originate.

2.2a Structure of the Atom

Atoms consist of a positively charged nucleus and negatively charged electrons surrounding the nucleus (Figure 2-1). The nucleus is a tiny ball of protons and neutrons and is

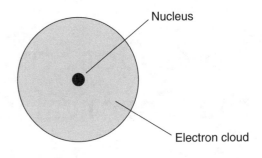

Figure 2-1 General structure of the atom. The nucleus at the center, composed of protons and neutrons, is surrounded by a cloud of electrons.

Nucleus

Electron cloud

much smaller than the space occupied by the electrons. That is, *the size of an atom is defined by the space taken up by the electrons.*

Table 2-1 lists the mass and charge of each of these elementary particles. Notice that the mass of the electron is quite small compared to the mass of a proton or neutron and can essentially be ignored for all practical purposes in organic chemistry.

An **atom,** strictly defined, has *no charge.* Therefore, in an atom, *the number of protons equals the number of electrons.* The number of protons that exist in the nucleus defines the atomic number and the element (e.g., a nucleus that has six protons has an atomic number of six and will always be defined as a carbon nucleus). If electrons are added or removed, the entire **species** (i.e., collection protons, neutrons, and electrons) then bears a charge and is therefore an **ion.** A negatively charged ion, or an **anion,** results from an excess of electrons. A positively charged ion, or a **cation** (pronounced *cat-ion,* not *cay-shun*), results from a deficiency of electrons. For example, an atomic ion that has six protons and five electrons is a cation of the carbon atom, because its atomic number is six. Its charge is +1.

All of the electrons surrounding the nucleus of an atom or ion reside in **orbitals.** We think of each orbital as a room, centered at the nucleus. Their specific purpose is to hold electrons—up to two electrons per orbital. Orbitals have various shapes, depending upon the orbital type, labeled *s, p, d,* etc. (this is discussed in Chapter 3 in the section on molecular orbital theory).

Each orbital is assigned to a particular **shell** of the atom. The only orbital in the first shell is an s orbital, called the 1s orbital (the "1" designates the shell we are talking about). In the

Table 2-1

Particle	Charge (a.u.)	Mass (a.u.)
Proton	+1	About 1
Neutron	0	About 1
Electron	−1	About 1/2000

second shell, there are four orbitals—an s orbital and three p orbitals, called "2s," "2p$_x$," "2p$_y$," and "2p$_z$." All three of these p orbitals are the same in every respect, except for their orientation (see Chapter 3). In the third shell, we find nine orbitals—one s orbital (3s), three p orbitals (3p$_x$, 3p$_y$, and 3p$_z$), and five d orbitals (we will focus on s and p orbitals only, because atoms in the second row of the periodic table, such as carbon, are composed of electrons occupying only those orbitals).

The main thing you need to understand about shells is that *the shell number correlates with the size of the orbital. The higher the shell number, the larger the orbital.* With a larger orbital, electrons have more room to roam around, and they reside, on average, farther from the nucleus. For example, when we compare s orbitals in the three different shells, we know that the size increases in the order 1s < 2s < 3s. Also, the 2p is smaller in size than the 3p orbital.

2.2b Electron Configuration

Electron configuration is a description of where electrons reside in atoms and molecules. It tells us specifically the number of electrons in every orbital. In determining the **ground state** (or most stable) electron configuration of an atom or ion, you need only know the relative energies of the various orbitals (discussed later) and follow three simple rules of quantum mechanics:

  1 There may be 0, 1, or 2 electrons in an orbital.
  2 Fill the lowest energy orbitals first.
  3 Don't pair the electrons unless forced to.

The first rule is a result of the possible spins an electron can have: either $+^1/_2$ (spin up) or $-^1/_2$ (spin down). When the first electron is placed in an orbital, it can be either spin—we are assuming that there is no preference. The second electron placed in an orbital, however, must be opposite in spin to the first electron, making the two electrons "spin-paired." *A third electron is not allowed in an orbital, because no two electrons may be in the same orbital with the same spin.*

The second rule is a summary of what's called the **aufbau principle.** Continually putting the next electron in the lowest energy orbital available ensures that we end up with all of the electrons arranged in the best (most stable) electron configuration. It is analogous to filling up a hotel beginning with the rooms on the ground floor and working upward.

The third rule is one of **Hund's rules.** The reason that electrons are not paired up unless forced is because of electron repulsion. Each electron is negatively charged, and negative charges repel one another. If there are two orbitals of equal energy in which to put the next electron—one that already has an electron in it and one that is empty—then it is better to put that electron in an empty orbital.

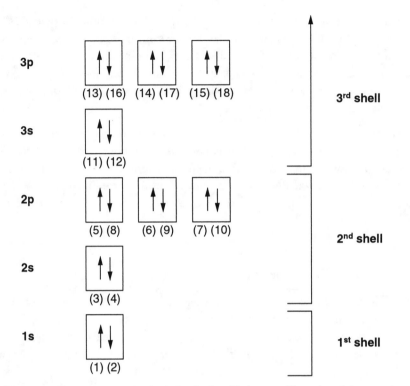

Figure 2-2 Energy diagram of atomic orbitals for the first 18 electrons. The order of electron placement is indicated in parentheses. Each successive electron is placed in the lowest energy available orbital. Notice that in the 2p and 3p sets of orbitals, no electrons are paired up until the addition of the fourth electron.

The relative energies of the common orbitals encountered in organic chemistry are shown in Figure 2-2. The 1s orbital (that is, the s orbital in the first shell) is the only orbital in the first shell and is the lowest in energy. Next comes the s orbital in the second shell (2s), and then the three 2p orbitals ($2p_x$, $2p_y$, and $2p_z$). In the third shell, the 3s orbital comes first, and then the three 3p orbitals. There are also five d orbitals found in the third shell, but (as mentioned earlier) they are not invoked much. Recall from general chemistry that the 4s orbital is actually lower in energy than the 3d orbitals, but this is not of concern here.

Following the three rules from quantum mechanics, the first 18 electrons are filled as indicated in Figure 2-2. The six electrons found in a carbon atom, for example, would fill the orbitals as shown in Figure 2-3, in which there are two electrons in the 1s orbital, two electrons in the 2s orbital, and one electron in each of two different 2p orbitals (it doesn't matter which two). The shorthand notation for this electron configuration is: $1s^2 2s^2 2p^2$, where the superscripts indicate the number of electrons in each type of orbital.

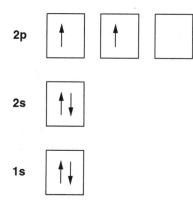

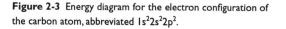

Figure 2-3 Energy diagram for the electron configuration of the carbon atom, abbreviated $1s^22s^22p^2$.

2.3 LEWIS DOT STRUCTURES

Lewis dot structures are the most convenient way in which chemists represent atoms, molecules, and ions. Lewis structures provide information about electrons in the species, and they also provide information about bonding—especially which atoms are bonded together and by what types of bonds. Before we talk about using Lewis structures in molecules, however, we begin with the representation of atoms.

2.3a Atoms

Lewis structures are concerned only with **valence electrons,** which are *electrons in the outermost (or valence) shell of an atom.* Valence electrons are distinguished from **core electrons,** which are all of the electrons occupying the shells that are not the valence shell—that is, the inner shells. The valence electrons are the only ones shown because the *valence electrons are exposed to other atoms and molecules, and they are the ones that are ultimately involved in bonding and chemical reactivity.*

The nucleus of each atom is represented by the element corresponding to the atomic number (the number of protons). A nucleus that contains seven protons, for example, is the nucleus of a nitrogen atom and would be represented by N. A chlorine nucleus that contains 17 protons would be represented by Cl.

The Lewis structures for the carbon and chlorine atoms are shown in Figure 2-4. For the carbon atom (Figure 2-3), we already know that there are two electrons in the first shell and four in the second. There are therefore two core electrons and four valence electrons. The four valence electrons are represented as dots around the C nucleus. For the chlorine

$$\cdot \overset{\displaystyle \cdot}{\underset{\displaystyle \cdot}{C}} \cdot \qquad : \overset{\displaystyle \cdot \cdot}{\underset{\displaystyle \cdot \cdot}{Cl}} \cdot$$

Figure 2-4 Lewis structure representation of the carbon atom (left) and the chlorine atom (right).

atom, there are 17 electrons total—2 in the first shell, 8 in the second shell, and 7 in the third shell, making 10 core electrons and seven valence electrons (Figure 2-2). The seven valence electrons are represented as the seven dots around the Cl nucleus, and the pairing of the electrons reflects the first rule from quantum mechanics (previously described).

If we examine the electron configurations of several atoms, we see that *atoms in the same column (or group) have the same number of valence electrons*. Furthermore, *the number of valence electrons an atom has is the same as its group number*. For example, the electron configuration of the fluorine atom is $1s^2 2s^2 2p^5$. It therefore has seven valence electrons, the same as the chlorine atom, and both atoms are found in the column designated as Group 7A elements.

2.3b Molecules

The primary difference between Lewis structures for atoms and molecules is that some electrons in molecules are involved in bonding, and the Lewis structures must show this. In particular, the type of bonding shown explicitly by Lewis structures is **covalent bonding**, which is *the sharing of a pair of electrons between two separate atoms* (discussed in greater detail in Chapter 3). To begin to understand covalent bonding, we make note of the peculiarity of quantum mechanics that *atoms are especially stable ("happy") when they have completely filled valence shells*. From Figure 2-2, we see that this means two electrons in the first shell (the so-called **duet**), and eight electrons in the second shell (the so-called **octet**). As evidence of this stability, note that the element with a completely filled first shell (two total electrons) is helium, and that with a completely filled second shell (10 total electrons) is neon. Both of these elements are part of the family of elements called the **noble gases,** which are highly unreactive.

The first molecule we examine is the simplest—the H_2 molecule. Each isolated hydrogen atom (H) has only one electron, which is one short of a complete valence shell (what it desires). If two hydrogen atoms are in close proximity, then the two total electrons can be shared between the two hydrogen nuclei, giving rise to a single covalent bond (the sharing of a pair of electrons between two nuclei). A covalent bond can be represented by a dash (X—Y) or a pair of adjacent dots (X:Y); in this book, we will use dashes. With the sharing of that pair of electrons, each hydrogen nucleus is surrounded by two electrons, and therefore "feels" that it has a complete valence shell. Figure 2-5a illustrates this point.

In the case of carbon (Figure 2-4), an isolated atom has four valence electrons, which is four less than an octet. In order to feel that it has an octet, the carbon atom needs to be surrounded by another four electrons, which can be accomplished in several ways. One way is to form four single bonds with hydrogen atoms (Figure 2-5b), making a molecule of CH_4 (methane). Each hydrogen atom that the carbon bonds to provides a share of one additional electron. Another way for the carbon atom to feel that it is surrounded by eight total

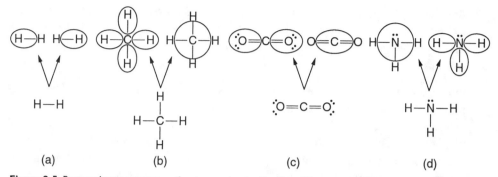

Figure 2-5 Bottom: Lewis structures of various molecules. Top: Two different ways of assigning the valence electrons to atoms for each Lewis structure. The ovals represent the different electron assignments. (a) H_2. Top left: The left-hand H atom has a complete valence shell of two electrons. Top right: The right-hand H atom has a complete valence shell. (b) CH_4. Top left: Each H atom is surrounded by two electrons. Top right: The C atom is simultaneously surrounded by eight electrons. (c) CO_2. Top left: Each O atom is surrounded by eight electrons. Top right: The C atom is surrounded by eight electrons. (d) NH_3. The N is surrounded by eight electrons, which includes a lone pair. Top right: Each H atom is surrounded by two electrons.

electrons in its valence shell is to form two double bonds with oxygen atoms, yielding a molecule of CO_2 (Figure 2-5c). Each oxygen atom provides a share of an additional two electrons.

The isolated nitrogen atom has five valence electrons and would therefore need three more electrons to fulfill its octet. NH_3 is one molecule that allows nitrogen to be surrounded by a total of eight electrons (Figure 2-5d). In forming this molecule, nitrogen has formed three bonds to hydrogen atoms, leaving two electrons not bonded. They remain as a **lone pair** of electrons on the nitrogen atom.

Recall from general chemistry the systematic procedure used to construct a Lewis structure, given the **connectivity** of the molecule (i.e., which atoms are covalently bonded together) and its total charge:

1 Count the total number of valence electrons.
  a The number of valence electrons contributed by each neutral atom is the same as its group number (e.g., $C = 4$, $N = 5$, $O = 6$, $F = 7$).
  b If the total charge is -1, -2, or -3, ADD 1, 2, or 3 valence electrons, respectively.
  c If the total charge is $+1$, $+2$, or $+3$, SUBTRACT 1, 2, or 3 valence electrons, respectively.
2 Write the skeleton of the molecule, showing only the atoms and a single covalent bond connecting each pair of atoms that you know must be bonded together.
3 Subtract two electrons for each single covalent bond drawn in step 2.
4 Distribute the remaining electrons as lone pairs around the atoms.
  a Start from the outside atoms and work inward.
  b Attempt to achieve an octet/duet on each atom.
5 If an atom is deficient of its octet, convert lone pairs from *neighboring* atoms into bonding pairs of electrons, thereby creating double and/or triple bonds.

Note that in step 5, the atom whose lone pair is being converted into bonding electrons does not actually lose its share of those electrons. Therefore, if that atom initially has an octet, it keeps its octet. The octet-deficient atom gains a share of an additional two electrons.

Applying the above five steps, let's construct a Lewis structure of NO_2^-, where N is the central atom. To satisfy step 1, we count 5 valence electrons for N, and 6 valence electrons for each O, for a total of 17. We add 1 electron for the overall −1 charge, for a total of 18 valence electrons. Step 2 has us write down the skeleton of the species. Because N is the central atom, each O must be bonded to the N, giving us O—N—O as the skeleton. The two necessary covalent bonds account for 4 valence electrons, leaving another 14 that must still be shown. This takes care of step 3. In step 4, we distribute those 14 electrons around the molecule, beginning from the outer atoms and working inward, with the intent of fulfilling each atom's octet as we go. Therefore, we can place 6 of those 14 electrons as 3 lone pairs on the O atom on the left. Those 3 lone pairs, plus the single covalent bond, give that O atom its octet. That leaves 8 electrons yet to be accounted for. We do the same with the O atom on the right, which takes care of another 6 valence electrons. That leaves two valence electrons not yet shown. The only place remaining for them is on the N atom. We have now completed step 4, and the resulting structure is shown on the left in Figure 2-6.

Currently, there is one atom that does not have its octet—the N atom. It has a single covalent bond to each of the two O atoms (for a share of four electrons), and a lone pair (for another two electrons), such that the N atom is surrounded by a total of six valence electrons. Step 5 suggests that we convert a lone pair of electrons from a neighboring O atom into an additional covalent bond between that O and the N atom. This is shown in Figure 2-6, and the resulting structure shows that the N atom has a double bond (for a share of four electrons), a single bond (for a share of two electrons), and a lone pair (for another two electrons). Eight total electrons now surround the N atom, giving it an octet. Additionally, the double bond on one O gives that O a share of four electrons, and the two lone pairs give it an additional four electrons so that it has an octet. And the O atom with three lone pairs did not change during the conversion—its three lone pairs give it six valence electrons, and the single covalent bond gives it a share of another two electrons for a total of eight.

If you continue to play with constructing Lewis structures, you will notice that each type of atom prefers to have a very specific number of bonds and a very specific number of lone pairs of electrons. Table 2-2 summarizes these preferences. Understanding this, it often becomes quite fast and easy to construct Lewis structures of molecules, if you know exactly

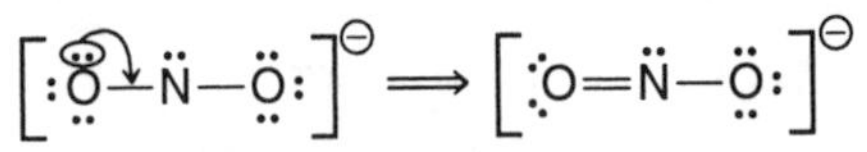

Figure 2-6 (Left) The result of the first four steps of constructing a Lewis structure. Notice that the N atom does not have an octet. (Right) Completion of the last step. The N atom is now surrounded by eight electrons.

$$\left[:\ddot{O} \overset{\frown}{\longrightarrow} \ddot{N} - \ddot{O}: \right]^{\ominus} \implies \left[:O = \ddot{N} - \ddot{O}: \right]^{\ominus}$$

Table 2-2

Atom	Preferred # of Bonds	Preferred # of Lone Pairs
H	1	0
C	4	0
N	3	1
O	2	2
F	1	3
Ne	0	4

which atoms are bonded to which (the connectivity). For example, in a molecule of HCN, where carbon is the central atom, you must try to find a Lewis structure in which the H atom has one bond and zero lone pairs, the carbon atom has four bonds and zero lone pairs, and the N atom has three bonds and one lone pair. This is easily obtained with H—C≡N: which has a triple bond between the carbon and nitrogen atoms.

One last thing pertaining to Lewis structures. For convenience sake, lone pairs of electrons are often not shown explicitly. Unless you are provided enough information to suggest otherwise, you may assume that all atoms will have their octets. Therefore, if an O atom has only one bond to it, and no lone pairs are shown, you may assume that the remaining six electrons needed to fulfill its octet are present in the form of three lone pairs. Similarly, if you see an N atom that has only a triple bond, you may assume that the remaining two electrons are present as a single lone pair.

As you will see throughout the next few sections, it is often important to keep track of ALL valence electrons. When you encounter Lewis structures in which lone pairs have been omitted, it is best to explicitly draw in those lone pairs, at least in the beginning. As you become more accustomed to working with Lewis structures, this will be less and less crucial. For now, don't cheat yourself—draw in those lone pairs explicitly!

2.4 FORMAL CHARGE

Within a molecule or ion, certain atoms may bear a positive or negative charge, while others may not. Such **formal charge** on each atom is quite important in understanding the reactivity of molecules, as you will see throughout the rest of this book. Specifically, knowing how to calculate formal charge is important in being able to apply one of the concepts introduced in a later chapter: "Charge is bad."

Most traditional textbooks suggest that you use an equation to calculate the formal charge. I recommend NOT using an equation, because it is simply one more thing that you must memorize; as I stressed in Chapter 1, *keep memorization to a minimum*. Instead, all we must do is apply what we already know about atoms: An atom has no charge if it has the same number of protons and electrons; it carries excess negative charge if it has more electrons than protons, and a positive charge if it has fewer electrons. The rest is simply *assigning* electrons to each atom in a covalent molecular species, and recognizing how many electrons each atom requires to achieve an overall 0 charge.

Let's begin with a C atom, which has an atomic number of six. The nucleus has six protons, and in order for there to be no charge, it must have six total electrons. This is the case with an isolated neutral C atom. Two of those electrons are in the $n = 1$ shell (the 1s orbital), and the remaining four are in the $n = 2$ shell (the 2s and 2p orbitals), which is the valence shell (Figure 2-3). In other words, when a C atom formally has four valence shell electrons, its formal charge is 0. Similarly, the isolated N, O, and F atoms all have a formal charge of 0, and have five, six, and seven valence shell electrons, respectively. Furthermore, all atoms in the same column require the same number of valence electrons to achieve a 0 formal charge. For example, Cl and Br atoms have a 0 formal charge when they formally have seven valence electrons, the same as F.

Assigning the number of valence electrons each atom has in a Lewis structure is straightforward. It involves two rules, which depend on whether those electrons exist as lone pairs or as bonding pairs: (1) A lone pair of electrons on an atom is assigned to that atom, and (2) for each bonding pair of electrons between two atoms, one electron is assigned to one atom, and the other electron is assigned to the other atom.

That's all you need to know to calculate the formal charge of any atom in a covalent compound. First, let's determine the formal charge on each atom in CH_4. We already know that in order for the C atom to have a formal charge of 0, it must possess four valence electrons, the same as in the isolated neutral atom. In the Lewis structure of CH_4, the C atom is bonded to each H atom by a single covalent bond, for a total of four bonds, and it has no lone pairs of electrons. Each single bond is comprised of two electrons, and one of those electrons is assigned to C, while the other is assigned to H. Since there are four total bonds on C, the C atom is assigned a total of four valence electrons. Therefore, the number of valence electrons assigned to the C atom is equal to the number it has in the isolated neutral. That is, the formal charge on that C atom is 0.

Similarly, each H atom has exactly one covalent bond, or one pair of bonding electrons. One of those electrons is assigned to the H. This is the same number of electrons as in the isolated neutral H atom. In CH_4, then, each H atom has a formal charge of 0.

The same strategy works for molecular ions, such as $[CN]^-$ (the negative charge is placed outside of the brackets to avoid indicating anything about which atom it resides on). The Lewis structure for this ion is :C≡N: (you should be able to come up with this from the five steps provided earlier), from which we can determine each atom's formal charge. We know that the isolated neutral N atom has five valence electrons. In this species, N has a lone pair and a triple bond. Both electrons of the lone pair are assigned to the N. In the triple bond, there are six electrons total, three of which are assigned to the N atom, and three of which are assigned to the C. In all, the N atom is assigned five valence electrons, the same as in the isolated neutral N atom. The formal charge on N is therefore 0.

As we saw before, the isolated neutral C atom has four valence electrons. In this molecule, however, C is assigned both of the electrons from the lone pair, and three electrons from the triple bond, giving it a total of five valence electrons. The C atom is therefore assigned one electron more than in the isolated neutral C atom, giving it a formal charge of −1.

A second way we could have deduced that −1 is the formal charge on C is to calculate by difference using the following rule: *The sum of the formal charges on all the atoms in the species must equal the total charge on the species.* In this case, the total charge is −1, which should equal the sum of the formal charges on C (i.e., FC(C)) and on N (i.e., FC(N)). We had already identified that FC(N) = 0, such that −1 = FC(C) + 0, or FC(C) = −1.

2.5 OXIDATION STATE

As with formal charge, many textbooks have you memorize an equation to calculate **oxidation state** (or **oxidation number**). That equation is slightly different from the one for formal charge, meaning that you will have two equations to memorize for these simple calculations. However, this is unnecessary, since oxidation state overlaps formal charge considerably. Both represent, in some way, the number of electrons on an atom relative to the number of protons; excess electrons give rise to negative oxidation states, and a deficiency of electrons gives rise to positive values for oxidation states.

In general, though, the oxidation state of an atom is different from the formal charge on that atom. The reason is that the way electrons are assigned in determining oxidation state is different from the method we used for formal charge. Specifically, the bonding electrons are assigned differently. To calculate formal charge, we saw that one electron from a bonding pair is assigned to one atom, and the other electron is assigned to the other atom. *To calculate oxidation state, both electrons from a pair of bonding electrons are assigned to the atom that is more electronegative* (recall from general chemistry the periodic table trend for electronegativity of atoms: Electronegativity increases from left to right across the periodic table and decreases down the periodic table). If the two atoms are identical, then their electronegativities are the same, and the pair of electrons is split evenly— one electron is assigned to each atom. Aside from the way in which we assign the bonding pairs of electrons, the procedures for calculating oxidation state and formal charge are identical.

To put this into practice, let's examine the same species we examined before, for which we calculated formal charge on each atom. First, look at CH_4, in which the only type of bond is a C—H single bond. The electronegativity of C is slightly greater than that of H (2.5 compared to 2.2). Therefore, for each of those single bonds, the C atom is assigned both electrons, leaving the H with none. The isolated neutral H atom has one electron, such that in calculating oxidation state, the H atom in CH_4 is deficient one electron. Therefore, each H atom's oxidation state is +1. The C atom, on the other hand, is assigned a total of eight valence electrons—both electrons from each of four bonding pairs. As we saw earlier, the isolated neutral C atom has only four valence electrons, leaving the C atom in CH_4 with four extra electrons. Its oxidation state is therefore −4.

Another way that we could have calculated the oxidation state on C is by knowing that *the sum of the oxidation states on all the atoms must be equal to the net charge on the species*. In this case, the oxidation state of the C atom (OS(C)) and the oxidation states of the four H atoms (OS(H)) must sum to 0. That is, $OS(C) + 4[OS(H)] = 0$. Solving this, we find that $OS(C) = -4$.

Similarly, we can calculate the oxidation state of each atom in $[:C\equiv N:]^-$. The only bond that exists is a triple bond between C and N. Since N is more electronegative than C, N is assigned all six of those bonding electrons. The N atom is therefore assigned eight valence electrons, compared to only five in the isolated neutral N atom. The N atom in $[CN]^-$ therefore has an oxidation state of −3. The C, on the other hand, is assigned only the lone pair of electrons, for a total of two valence electrons, which is two less than in the isolated neutral C atom. Its oxidation state is therefore +2. As you can see, the sum of the oxidation states is equal to the total charge of −1.

Table 2-3 lists the oxidation states and formal charges on each atom in the two species we have examined. Comparing the results of the two types of calculations helps to make clear their similarities and differences. Particularly, in both types of calculations, the number of electrons in the isolated neutral atom does not change; it is the number of electrons each atom is assigned that is different.

Table 2-3

Species	Atom	Valence Electrons in the Isolated Neutal Atom	Valence Electrons Assigned		Formal Charge	Oxidation State
			F.C.	O.S.		
CH_4	C	4	4	8	0	−4
	H	1	1	0	0	+1
CN^-	C	4	5	2	−1	+2
	N	5	5	8	0	−3

2.6 RESONANCE

Resonance is a phenomenon that occurs within a species that has *two or more valid Lewis structures.* Each valid Lewis structure is called a **resonance structure,** or a **resonance contributor.** A good example is with NO_2^-. Earlier we had derived a valid Lewis structure by going through the five steps. Note the situation we had after the completion of the first four steps (Figure 2-6). The species was symmetric, with three lone pairs and a single covalent bond on each O atom, and one lone pair and two single bonds on the N atom. The Lewis structure was completed with step 5 by converting one lone pair of electrons from the O atom on the left into a bonding pair between the O and N atoms. However, in doing so, we could have used a lone pair from either O atom—both are equally good. If a lone pair is used from the O atom on the left, the Lewis structure in Figure 2-6 is obtained. On the other hand, if the lone pair comes from the O atom on the right, a second Lewis structure is obtained. It therefore appears that there are two equally good Lewis structures for NO_2^-, as shown in Figure 2-7. In other words, there are two resonance contributors of NO_2^-.

Although there are two resonance contributors that can be drawn for NO_2^-, the NO_2^- species has *only one* definite structure. Experimentally we know that NO_2^- has two of the exact same nitrogen-oxygen bonds, and each one behaves as something intermediate between a single and a double bond. We also know from experiment that each oxygen atom is identical and carries a partial negative charge—that is, a charge that is somewhat less than a full -1 charge.

Each of the two resonance forms of NO_2^- appears to disagree substantially with what we know to be true about the real species. For example, in each resonance contributor, there are two different bonds—one single bond and one double bond. There are also two different oxygen atoms—one that has no charge, and one whose charge is -1. In other words, NO_2^- appears to be a species that cannot be described accurately by the rules of the Lewis structures we learned—Lewis structures have limitations! As it turns out, any species that has two or more valid Lewis structures—that is, has resonance—suffers from the same problem.

How do we reconcile the difference between the Lewis structure of a species like NO_2^- and its actual structure? The answer is quite simple—*we take each resonance contributor to be an imaginary species, and we take the actual structure to behave as something like the average of*

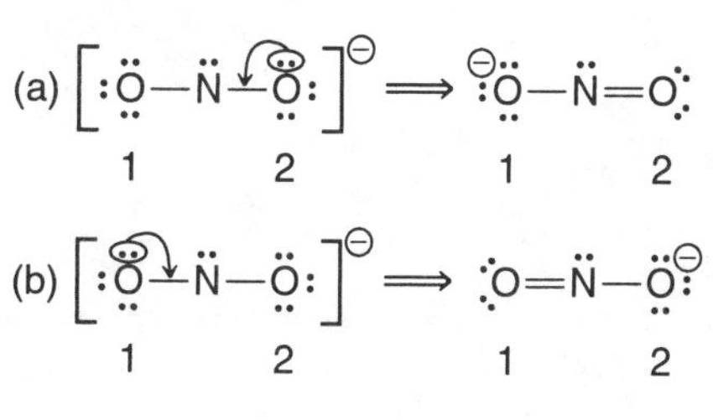

Figure 2-7 Derivation of the two resonance contributors of NO_2^-. The oxygen atoms are labeled to emphasize that they are different atoms. (a) Conversion of a lone pair of electrons from the O on the right leads to one resonance structure. (b) Conversion of a lone pair from the O on the left leads to the other.

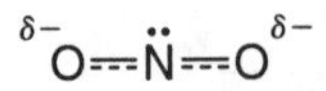

$$\overset{\delta-}{O} = \overset{\cdot\cdot}{N} = \overset{\delta-}{O}$$

Figure 2-8 The real structure, or resonance hybrid, is roughly the average of the individual resonance contributors. Each dashed line in the structure on the right indicates a partial bond, and each δ^- indicates a partial negative charge.

all of its resonance contributors. In other words, the one, true species is a **resonance hybrid of its resonance contributors.** In Figure 2-7, one resonance contributor shows that there is a double bond between N and O(1), whereas the other resonance contributor shows a single bond between those two atoms. The resonance hybrid should reflect something like the average between a single bond and a double bond, or about 1.5 bonds. Likewise, in one resonance contributor, there is a double bond between the N and O(2), whereas the other resonance structure shows that there is a single bond there. Again, in the resonance hybrid, there is something like 1.5 bonds between N and O(2). This is indicated in Figure 2-8 by a solid bond plus a dashed bond between the N and each O.

The same analysis can be done with the formal charge. In one resonance contributor (Figure 2-7), O(1) has a formal charge of −1, and O(2) has no formal charge. In the other resonance contributor, it is the reverse—O(1) has no formal charge, but O(2) has a formal charge of −1. In the resonance hybrid, the charge on O(1) is the average of the two, which is −0.5. By the same token, the charge on O(2) should be about −0.5. In Figure 2-8, this is indicated by a "δ^-" on each O, which stands for "partial negative charge."

In order to arrive at any resonance hybrid, we must be able to draw all of its resonance contributors and then take their average. To draw all resonance contributors (as we learn to do in Section 2.9), we should understand the relationship between any pair of them. *Resonance contributors are related by the rearrangement of electrons, while the atoms themselves remain frozen.* This can be seen for the resonance contributors of NO_2^-, given that each one is the result of having moved a different pair of electrons in step 5 of the Lewis structure rules.

As a consequence of their relationship, resonance contributors can be interconverted by the simple movement of electrons. To convert one resonance form of NO_2^- into the other (Figure 2-9), a lone pair of electrons on O^- is converted into a covalent bond to form an N=O double bond. As a result, the formal charge on the O atom goes from −1 to 0. Simultaneously, a pair of electrons from the N=O double bond on the other side of the N atom is kicked over as a lone pair on the O atom that is initially neutral. The result is the conversion of the N=O bond to an N—O single bond, with the formal charge on that O atom going from 0 to −1.

$$\overset{\ominus}{\underset{\cdot\cdot}{O}} - \overset{\cdot\cdot}{N} = \overset{\cdot\cdot}{O} \longleftrightarrow \overset{\cdot\cdot}{O} = \overset{\cdot\cdot}{N} - \overset{\cdot\cdot}{\underset{\cdot\cdot}{O}}{}^{\ominus}$$

Figure 2-9 Conversion of one resonance contributor of NO_2^- into the other. A lone pair is converted into a bonding pair, while a second bonding pair is converted into a lone pair.

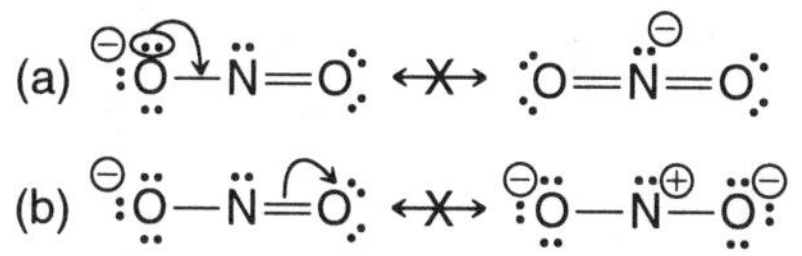

(a) $\overset{\ominus}{:}\overset{..}{O}\text{---}\overset{..}{N}=\overset{..}{O}: \longleftrightarrow\!\!\!\times\!\!\!\longrightarrow :O=\overset{\ominus}{\underset{}{N}}=\overset{..}{O}:$

(b) $\overset{\ominus}{:}\overset{..}{O}\text{---}\overset{..}{N}\overset{..}{=}\overset{..}{O}: \longleftrightarrow\!\!\!\times\!\!\!\longrightarrow \overset{\ominus}{:}\overset{..}{O}\text{---}\overset{\oplus}{N}\overset{..}{\underset{..}{O}}\overset{\ominus}{:}$

Figure 2-10 Valid resonance structures of NO_2^- cannot be drawn with the movement of only one of the two pairs of electrons indicated in Figure 2-9. (a) The N in the resulting resonance structure is surrounded by 10 electrons, breaking the octet. (b) That N atom is surrounded by only 6 electrons.

Notice in Figure 2-9 that there are two types of arrows used when working with resonance contributors. One is the double-headed straight arrow ($\leftrightarrow$), which, when placed between two structures, indicates that they are resonance contributors of the same overall species. The other type of arrow is the curved, single-headed arrow ($\frown$), which shows the movement of a pair of electrons. In Figure 2-9, one curved arrow originates from a lone pair and points to the region between the N and O, which signifies the conversion of that lone pair of electrons into a covalent bond. The other curved arrow originates from the middle of the double bond and points to an O atom, which signifies the conversion of a pair of electrons from the double bond into a lone pair of electrons.

When drawing resonance contributors, it is important to remember that a resonance contributor, by definition, is a valid Lewis structure; we must therefore adhere to the octet rule. Common atoms encountered in organic chemistry, such as carbon, nitrogen, and oxygen, are not allowed to exceed their share of eight electrons (for hydrogen, it is two electrons). At the same time, atoms want to attain their octets. Therefore, *a valid resonance form should have as many atoms with octets as possible.*

In our example with NO_2^-, it is not appropriate to draw only one of the two curved arrows in Figure 2-9. If the curved arrow on the left was drawn without the curved arrow on the right, this would result in four bonds and one lone pair on the nitrogen atom, for a total share of ten electrons; that would therefore exceed the octet (Figure 2-10a). If, however, the curved arrow on the right was drawn without the one on the left, the result would be a nitrogen atom with a share of only six electrons (Figure 2-10b). This is not considered a viable resonance structure, since we know that it doesn't have the maximum number of atoms with their octet; there are two other resonance forms in which all three atoms have their octets (Figure 2-9).

2.7 APPLICATION: TRACKING FORMAL CHARGE

As we will see in Chapter 5, charge is one of the most important factors used to understand chemical reactions. Furthermore, during a chemical reaction, there may be several behind-the-scene steps, called the mechanism (see Chapter 6), where a formal charge is transferred from atom to atom. It is therefore necessary to be able to track formal charges throughout a mechanism. One way to do this is to stop after each step of the mechanism and calculate the formal charge on each atom in every molecule. As you might imagine, however, this can become tedious and time-consuming. Fortunately, it can be much faster and easier.

A more efficient way to keep track of charge is to realize that each step of a reaction involves the breaking or formation of bonds, which is usually nothing more than the inter-conversion of electrons between bonding pairs and lone pairs. If an atom is not directly involved with those electrons, then the number of electrons it is formally assigned does not change, and neither does its formal charge. On the other hand, if a lone pair on an atom is converted into a bonding pair, or vice versa, then the number of electrons it is assigned does change, and so will its formal charge. Specifically, *if a bonding pair is converted into a lone pair, then the number of electrons that atom is formally assigned increases by one, and its formal charge decreases by one. Likewise, if a lone pair is converted into a bonding pair, that atom formally loses one electron, and its formal charge increases by one.*

This idea can be applied to a variety of reactions. One of the simplest, but one that illustrates what we just discussed, is an acid-base reaction, such as that between HCl and [OH]⁻ (Figure 2-11). In the reactant species, the H atoms and the Cl all have a formal charge of 0, where-as the O atom has a formal charge of −1 (you should verify this). For the products, we could count the number of valence electrons each atom is formally assigned to determine the formal charge on each atom. The better way, as previously discussed, is to make note of what happens to the electrons when each bond is broken or formed. It appears that the bond between the H and the Cl is broken during the course of the reaction, and those elec-trons are converted into a lone pair on the Cl atom. Therefore, the Cl atom is formally assigned one additional electron in the products, and its formal charge decreases by one. Since it was initially 0, it goes to −1. In addition, a lone pair on the O atom is converted into an O—H bond. Therefore, the number of electrons the O atom is formally assigned decreases by one, and the formal charge on it increases by one, going from −1 to 0. And finally, the H atom that was initially bonded to the Cl atom ends up with the same number of bonds and lone pairs as it initially had. Its formal charge therefore does not change during the course of the reaction.

Another useful method is to quickly recognize those atoms in a molecule that bear a nonzero formal charge. To do this, you must quickly recognize scenarios in which an atom has no formal charge. Table 2-2, which provided the number of bonds and lone pairs that atoms of H, C, N, O, and F prefer, provides those scenarios that give rise to 0 formal charge on those atoms (verify this!). Whenever an atom is NOT found in those situations, you can be guaranteed that it will have a formal charge.

$$:\ddot{\underset{\cdot\cdot}{Cl}}-H + \overset{\ominus}{\underset{\cdot\cdot}{:O}}-H \longrightarrow :\overset{\ominus}{\underset{\cdot\cdot}{Cl}}: + H-\underset{\cdot\cdot}{\overset{\cdot\cdot}{O}}-H$$

Figure 2-11 Reaction of HCl with [OH]⁻. We can quickly see that, during the course of the reaction, the formal charge on Cl decreases by one, and that on O increases by one.

For example, an O atom has no formal charge when its octet is composed of two bonds and two lone pairs (Table 2-2)—in such a case, the O atom is formally assigned a total of six valence electrons, which is the same as the number of valence electrons in the isolated neutral O atom. If, however, an O atom is found in which its octet is composed of three bonds and one lone pair, you should be able to quickly recognize that it will NOT have a formal charge of 0. In this case, it turns out that the O atom is formally assigned five electrons, giving it a formal charge of +1.

Suppose, now, that an F atom is found in which its octet is comprised of four lone pairs of electrons. In order for it to have a formal charge of 0, an F atom must have three lone pairs and one bond. Therefore, an F atom that has four lone pairs must have a formal charge, and we calculate that formal charge to be −1—it is formally assigned eight valence electrons, but the isolated neutral atom has only seven.

To put everything together, let's look at another example in which we track formal charge in the first few steps (Figure 2-12) of the mechanism for what is called an "ester hydrolysis" reaction (a reaction you probably won't encounter until your second semester of organic chemistry). The complete mechanism is provided in Problem 2.8 at the end of this chapter.

There are no formal charges assigned in Figure 2-12, so before we can track a formal charge, we must first determine the formal charges that exist. In the molecule on the left in step 1, each H atom has one bond and no lone pairs. According to Table 2-2, each is formally assigned the same number of valence electrons as in the isolated neutral H atom, and is therefore assigned a formal charge of 0. Similarly, each C atom has four bonds and no

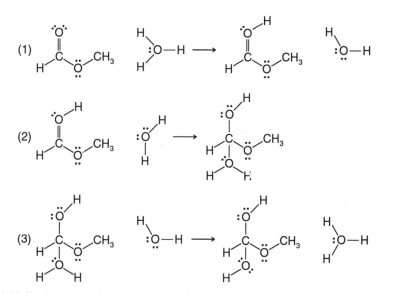

Figure 2-12 The first three steps of an ester hydrolysis reaction, in which formal charges are not assigned.

lone pairs, which, according to Table 2-2, gives them formal charges of 0 as well. Similarly, each O atom has two bonds and two lone pairs, for a formal charge of 0.

In the species on the right, all of the hydrogen atoms have one bond and no lone pairs, for a formal charge of 0. The oxygen atom in that species, however, is different from that given in Table 2-2. It has three bonds and one lone pair, whereas, we know from Table 2-2 that in order for it to have a formal charge of 0, it must have two bonds and two lone pairs. Its formal charge must therefore be calculated. We assign it a total of five valence electrons—one from each of the three bonds and two from the lone pair. In the isolated neutral O atom, however, there are six valence electrons. The O atom in the molecule is therefore deficient one electron, and we assign it a formal charge of +1, as shown in Figure 2-13.

During the course of the reaction in step 1, very little changes. Therefore, for most of the atoms, formal charge doesn't change. However, notice that the doubly bonded O atom on the reactant side appears to have one of its lone pairs of electrons converted to a single covalent bond. Whereas on the reactant side of that equation that oxygen atom is assigned both of those electrons, on the product side it is assigned only one of them. It therefore seems that that O atom loses one electron in going from the reactants to the products. Its formal charge must then increase by 1, going from 0 to +1. This is shown in Figure 2-13.

Similarly, the O atom in the H_3O species appears to have one of its covalent bonds converted into a lone pair. As a result, it is assigned one additional electron in the products than in the reactants, such that its formal charge decreases by 1. It goes from +1 to 0. Again, this is shown in Figure 2-13.

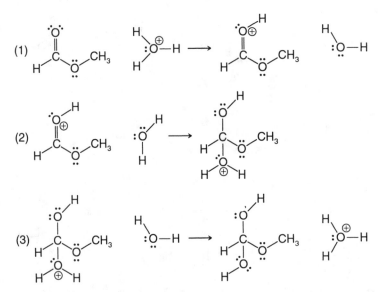

Figure 2-13 The first three steps of an ester hydrolysis reaction with the formal charges assigned.

In step 2, the reactant molecules are the same as that on the product side of step 1. We can therefore simply redraw the products from step 1, in which all atoms have no formal charge except the doubly bonded O.

During the course of the reaction in step 2, there are three atoms involved with bond breaking and bond forming. The O with the double bond has one of its bonds converted to a lone pair, so that it formally gains one electron, and its formal charge goes from +1 to 0. The carbon atom with the double bond appears to lose a bond on one side and gain a bond on the other. Overall, the situation is the same on the reactant and product side—four bonds and no lone pairs—so that its formal charge remains 0. Finally, the O atom on H_2O has one of its lone pairs converted into a bond; it formally loses one electron, and its formal charge goes from 0 to +1. These are all summarized in Figure 2-13.

In step 3, we have seen both reactant species before. The larger species can be redrawn from the products in step 2, where there is a +1 formal charge on the bottommost O atom. H_2O was seen in both step 1 and in step 2, where the formal charge on all three atoms was 0. In going from reactants to products, that bottommost O atom has a bond converted to a lone pair, so that its formal charge goes from +1 to 0. The O atom from H_2O, on the other hand, has a lone pair converted to a bond, and its formal charge increases from 0 to +1. Once again, these findings are summarized in Figure 2-13.

2.8 APPLICATION: IDENTIFYING REDOX REACTIONS, OXIDIZING AGENTS, AND REDUCING AGENTS

Reduction-oxidation (redox) reactions are a common and important class of reactions in organic chemistry. A reaction is a redox reaction if there is a transfer of electrons from one reactant species to another. One species is reduced, while another is simultaneously oxidized. A species is **reduced** if an atom in that species gains electrons (its oxidation state is decreased). A species is **oxidized** if an atom in that species loses electrons (its oxidation state is increased). Therefore, identifying a redox reaction simply involves identifying whether there has been a *transfer of electrons* between separate reactant species, *according to the rules in which we assign electrons for oxidation states.*

Such a transfer of electrons occurs when the relative electronegativities of atoms involved in a bond are different in the reactants than in the products (recall the periodic table trend—electronegativity increases up and to the right). One example of this is the substitution reaction shown in Figure 2-14, in which the Cl atom in CH_3Cl is replaced by a CH_3 group, producing $H_3C—CH_3$. In the reactant CH_3Cl, the carbon atom is less electronegative than the chlorine atom to which it is bonded. Both electrons in the carbon-chlorine bond are therefore assigned to the chlorine atom, according to how we determine oxidation state. In the products, that carbon atom is bonded to another carbon atom of

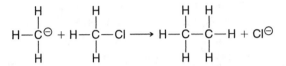

Figure 2-14 The carbon atom of CH$_3$Cl is reduced. Initially it is bonded to a Cl atom, which is more electronegative than C. The Cl atom is therefore assigned both electrons of the carbon-chlorine bond. In the products, however, that C is bonded to another C, such that the electrons in the carbon-carbon bond are equally shared. During the course of the reaction, then, the carbon atom of CH$_3$Cl appears to gain one electron.

equal electronegativity, so that the pair of bonding electrons is equally shared—one electron from the bond is assigned to each C atom. During the course of this reaction, then, the carbon atom in the reactants appears to gain one electron, and, as a result, is reduced.

In a different substitution reaction (Figure 2-15), a molecule of CH$_3$Cl is transformed into CH$_3$Br. As a result, the carbon-chlorine bond becomes a carbon-bromine bond. Because both chlorine and bromine are more electronegative than carbon, the relative electronegativities remain unchanged. Consequently, the assignment of electrons also remains unchanged, and there is no oxidation or reduction that takes place. A substitution reaction like this is therefore not considered a redox reaction.

Figure 2-16 provides an example of a redox reaction that involves a double bond. The carbon-carbon double bond in cyclohexene is converted into a single bond, and each of those carbon atoms gains a carbon-hydrogen single bond. Overall, then, each carbon atom loses a bond to the other carbon and gains a bond to hydrogen, which is less electronegative than carbon. In the carbon-carbon bond that is lost, each carbon atom has an equal share of the two electrons, or has one electron each. In the carbon-hydrogen bond, the carbon atom is assigned both electrons. As a result, each carbon atom gains one electron and is reduced in the process.

In addition to recognizing whether a redox reaction has taken place, we can also begin to make predictions as to whether a redox reaction *will* take place. If so, will it result in the oxidation or the reduction of the organic species (the species that is often of interest)? Making such predictions requires recognizing those species that tend to cause oxidation as well as those species that tend to cause reduction.

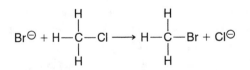

Figure 2-15 The C atom of CH$_3$Cl is neither oxidized nor reduced. The reason is simply that a Cl has been replaced by a Br, and both the Cl and Br atoms are more electronegative than C.

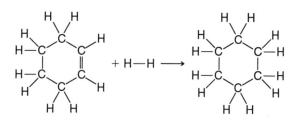

Figure 2-16 The molecule containing the carbon-carbon double bond is reduced. This is because each carbon of the double bond loses one bond to carbon (the other doubly bonded carbon atom) and gains a bond to hydrogen. Because carbon is more electronegative than hydrogen, each of the two carbon atoms gains one electron in the process.

An **oxidizing agent** is defined as that which causes another species to be oxidized in a reaction. It is recognized as the species that is itself reduced. In order to be reduced, *oxidizing agents tend to possess an atom (typically a metal) that is initially in a significantly positive oxidation state*. Common examples include $KMnO_4$, CrO_3, $H_2Cr_2O_7$, and OsO_4. In $KMnO_4$, an ionic compound composed of K^+ and $[MnO_4]^-$, the oxidation state on Mn is +7. In CrO_3, the oxidation state on the Cr is +6. The same is true for Cr in $H_2Cr_2O_7$. In OsO_4, the oxidation state of Os is +8.

Similarly, we can make generalizations about *reducing agents*. A **reducing agent** is that which is oxidized in a redox reaction. In contrast to oxidizing agents, reducing agents therefore tend to have metal atoms in low oxidation states. Common examples of reducing agents include $[AlH_4]^-$ (which is part of $LiAlH_4$) and $[BH_4]^-$ (which is part of $NaBH_4$). In $[AlH_4]^-$, the oxidation state on Al is +3. Similarly, the oxidation state on B in $[BH_4]^-$ is +3. Both of these oxidation states are considerably smaller than those we previously saw on the metal atoms in the common oxidizing agents. Other examples of reducing agents include elemental metals, such as Zn, which have an oxidation state of 0. Hydrogen gas, H_2, in the presence of a solid metal catalyst, is also a common reducing agent (Figure 2-16). Although not a metal, the hydrogen atoms of which H_2 is composed are in low oxidation states—in this case, their oxidation state is 0.

A second way in which to identify common oxidizing and reducing agents is in their hydrogen or oxygen content. Notice that *the oxidizing agents previously shown have a significant number of oxygen atoms;* they tend to donate highly electronegative oxygen atoms to organic species, thereby causing their oxidation. On the other hand, *reducing agents tend to have an abundance of hydrogen atoms.* An organic species is usually reduced if it gains hydrogen atoms.

With this knowledge of oxidizing and reducing agents, we can begin to make predictions about redox reactions. For example, we might be asked to predict the product of the following reaction:

$$H_3COH + KMnO_4 \rightarrow ?$$

given the following four choices:

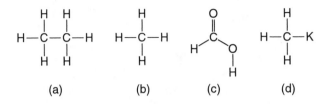

(a) (b) (c) (d)

Because $KMnO_4$ is an oxidizing agent (the metal atom is in a +7 oxidation state, and the species has plenty of O atoms), we should choose an answer that illustrates that the organic reactant molecule (H_3COH) has been oxidized. In CH_3OH, the carbon atom is singly bonded to three hydrogen atoms and one oxygen atom. Choice (c) is the only product molecule that has replaced a carbon-hydrogen bond with a bond involving an atom with greater electronegativity than carbon. Choice (c) is therefore the only molecule that is the product of an oxidation reaction.

To apply what we now know about reducing agents, let's predict the product of the following reaction from the given choices:

$$H_2C{=}O + LiAlH_4 \rightarrow ?$$

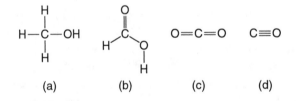

(a) (b) (c) (d)

Realizing that $LiAlH_4$ is a reducing agent (the metal atom is in a +3 oxidation state, and the species also contains several H atoms), we should choose the product for which there is an atom whose oxidation state is decreased from that in the reactant. In the reactant molecule, the carbon atom has two bonds to hydrogen and two bonds to oxygen. Of the given choices, choice (a) is the only molecule in which a carbon-oxygen bond has been replaced by a bond involving an atom less electronegative than carbon; it is therefore the only logical choice.

2.9 APPLICATION: DRAW ALL RESONANCE CONTRIBUTORS OF . . .

Being able to draw all valid resonance structures of a species is important for two reasons. First, it demonstrates a command of the resonance concept. Second, and perhaps more importantly, the number of resonance structures that can be drawn correlates with that species' stability—something that is critical to know in order to understand its reactivity.

All else being equal, *the more resonance structures that can be drawn, the more stable the species.* We touch more upon this idea in Chapter 5, and it will be developed more fully in your traditional textbook.

Drawing all resonance contributors can be straightforward if we use a systematic approach. We begin by applying what we already know about resonance contributors:

1. A resonance contributor must be a valid Lewis structure.
   a. If possible, all atoms should have an octet.
   b. Some atoms in the second row, like B, C, and N, may have less than the octet, but may not have more.
2. Resonance contributors are related by moving around electrons within the species, while the atoms remain frozen.

Examining Figure 2-9, which shows the conversion of one resonance contributor of NO_2^- into the second, provides insight into the systematic way of drawing all resonance contributors of a given species. Notice in Figure 2-9 which types of electrons are involved—lone pairs and pairs of electrons from double bonds. Specifically, a lone pair of electrons from one O atom is converted into a bonding pair, thereby converting a single bond into a double bond. Simultaneously, a pair of electrons from a different double bond is converted into a lone pair on another O atom, thereby converting that second double bond into a single bond.

In general, *the types of electrons that are involved in drawing different resonance contributors are lone pairs of electrons and pairs of electrons from multiple bonds*—double bonds and triple bonds. The reason is that, as we saw in the case of NO_2^-, if a pair of electrons from a double bond is shifted elsewhere, a single covalent bond remains. Similarly, if a pair of electrons from a triple bond is shifted elsewhere, a double bond still remains. However, if the pair of electrons from a single bond is shifted elsewhere, there are no electrons left to keep the atoms bonded together!

It will help to look at some examples that illustrate the most common ways in which electrons are shifted in going from one resonance contributor to another (Figure 2-17). First, let's look at the allyl anion, $CH_2CHCH_2^-$ (Figure 2-17a). We first start with any feasible Lewis structure obtained from the five steps we learned earlier. The result is a structure in which a central C atom forms a double bond to one C and a single bond with another C. To search for another resonance structure, we convert the lone pair of electrons into a bonding pair. To avoid five bonds to C (a share of 10 electrons), a pair of electrons from the double bond is kicked over to form a lone pair of electrons, thereby completing the new resonance contributor. A third resonance contributor does not exist. Convince yourself of this by keeping all atoms frozen in space, moving around only lone pairs and pairs of electrons from the double bond.

Next we examine the allyl cation (Figure 2-17b), in which the lone pair of electrons has been removed from the allyl anion. Notice that in this species, it is impossible for all of the

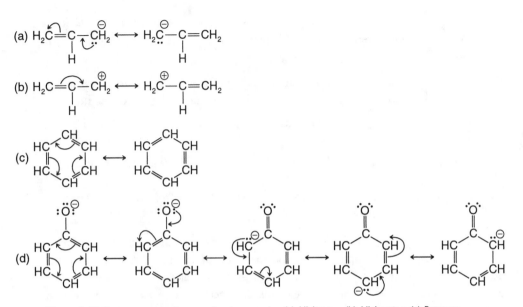

Figure 2-17 Resonance contributors of various species. (a) Allyl anion. (b) Allyl cation. (c) Benzene. (d) Phenoxide ion.

atoms to have an octet (prove this to yourself!)—one C atom has a share of only six electrons, and its formal charge is +1. Unlike the allyl anion, there are no lone pairs to convert into bonding pairs. However, a pair of bonding electrons from the double bond can still shift over to form a second bond between the central C and the C atom on the other side.

Figure 2-17c is a molecule of benzene, which is a six-membered ring of C atoms that has alternating single and double bonds. Again, there are no lone pairs, but a pair of electrons from a double bond can shift over to the other side, converting a single bond into a double bond. If only one of the curved arrows was drawn in Figure 2-17c, we would end up with a C atom that has five bonds. To avoid that situation, a second curved arrow is drawn, showing that a pair of electrons from another double bond is shifted to the other side. If only those two curved arrows were drawn, we would still end up with a C atom that has five bonds. The third arrow must therefore be drawn in. In all, there are six electrons that are shifted around to arrive at the new resonance contributor. That resonance contributor turns out to be the only other one, because if the electrons were shifted a second time, the result would be identical to the first.

Finally, let's draw all of the resonance contributors for the phenoxide ion (Figure 2-17d). In going from the first to the second resonance contributor, we perform the same six-electron movement as we did in benzene (Figure 2-17c). To go from the second to the third contributor, we convert a lone pair of electrons from the O atom into a second bond to the C. That forces a pair of electrons from a C═C double bond to be kicked over as a lone pair of electrons on C, so as to avoid five bonds and breaking the octet. The fourth and fifth resonance contributors are each the result of converting a lone pair into a bonding pair, with the simultaneous conversion of another bonding pair into a lone pair. There are a total of five resonance contributors for this species. Further movement of the lone pair of electrons on the C atom in the last resonance contributor would be a duplicate of one of the five that are shown.

Notice that the last three resonance contributors of phenoxide are each achieved by the same type of electron movement as in the interconversion of the NO_2^- contributors, as well as the allyl anion contributors. If you do not see this, then you will have a tendency to overlook some contributors.

Problems

2.1 For each of the four species in Figure 2-17, draw the resonance hybrid that most accurately depicts the species (as an example, see Figure 2-8).

2.2 Write the electron configuration of an isolated O atom. How many unpaired electrons are there? How many core electrons are there? How many valence electrons? Draw its Lewis structure.

2.3 From the number of bonds and lone pairs of electrons each atom prefers, complete the Lewis structure (including all lone pairs) of each of the following molecules with the given connectivity.

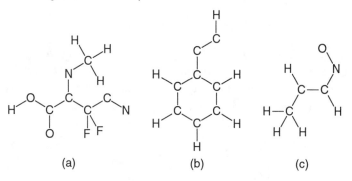

(a)   (b)   (c)

2.4 Draw the Lewis structure of each of the following molecules and calculate the formal charge of every atom. Include all lone pairs in the Lewis structures.
 a $[CH_3O]^-$ (C is the central atom)
 b HNO_3 (N is bonded to three O atoms, and the H is bonded to one O)
 c C_2H_6 (the two Cs are bonded together, and each C is bonded to three Hs)
 d C_2H_4 (the two Cs are bonded together, and each C is bonded to two Hs)
 e O_3 (a central O atom is bonded to two other Os)

2.5 Calculate the oxidation state on every atom in each of the following species.
 a–e Each species in Problem 2.4.
 f $AlCl_3$ (Al is the central atom)

g $B_2H_6{}^{2-}$ (one B is bonded to the second, and each B is also bonded to three Hs)
Note: The electronegativity of B is smaller than that of H
h MnO_2 (Mn is the central atom)

2.6 Using Table 2-2, quickly identify the atoms in the following species that will NOT have a formal charge of 0.

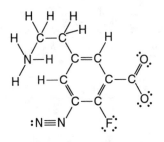

2.7 Draw all resonance contributors of the following species.
a O_3
b $HCO_2{}^-$ (C is the central atom)
c $CO_3{}^{2-}$ (C is the central atom)
d

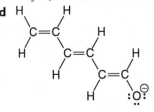

2.8 The following is a mechanism for an "ester hydrolysis" reaction. It is comprised of six individual steps, where an organic (carbon-containing) product of each step is a reactant in the next step. The first three steps have formal charges assigned. Assign the formal charge on every atom in the remaining three steps, using the method from Section 2.8 to track the formal charges.

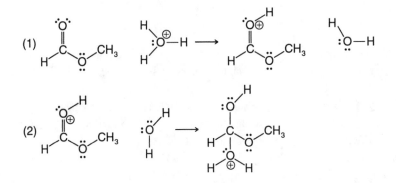

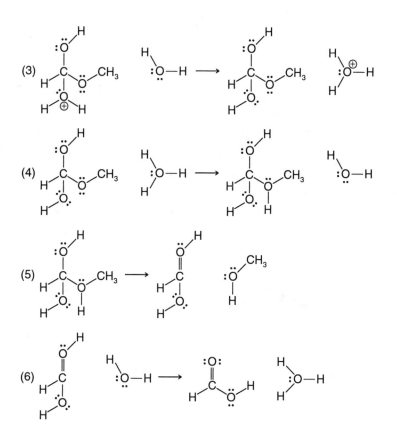

2.9 Predict the product of the following reaction. (Hint: Think about what kind of reagent Zn(s) is.)

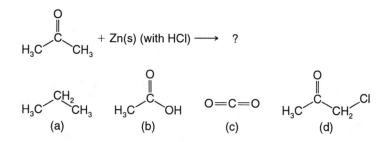

2.10 Which of the following pairs are NOT related by resonance (i.e., resonance contributors of the same overall species)? For each pair that is not, explain. (Note: all lone pairs of electrons might not be shown.)

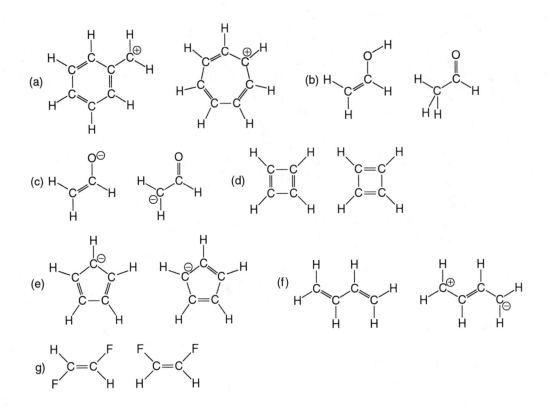

Molecular Geometry and Dipole Moments

3

3.1 INTRODUCTION

The specific shape of a molecule dictates a number of important properties. As indicated by the chapter title, one of those properties is the dipole moment, or polarity. Also, as you will see in Chapter 4, the three-dimensional nature of organic compounds gives rise to an interesting phenomenon, called *stereoisomerism*. Both stereoisomerism and polarity contribute to a given species' specific type of chemical reactivity.

Water is perhaps the best example with which to demonstrate the importance of molecular geometry. Water is essential for life, largely because of its unique physical and chemical behavior. Many of those properties, in turn, are a direct result of the fact that the water molecule, H_2O, has a strong dipole moment. From its Lewis structure, we know that the connectivity of the atoms is H—O—H. But the *Lewis structure does not tell us the specific shape*. As it turns out, the water molecule is bent, with an H—O—H angle of around 105°. If, instead, the molecule was linear (i.e., with an angle of 180°), then it would have no dipole moment whatsoever. Without a dipole moment, the boiling point of water would be much lower than 100°C, and it would be incapable of dissolving important species in the body, including the Na^+ and K^+ ions (see Chapter 7). Life would be very different.

3.2 VSEPR THEORY AND 3-D MOLECULAR GEOMETRY

Lewis structures are quite convenient in that they show which atoms are connected together by covalent bonds and by what type of covalent bond (single, double, or triple). However, *Lewis structures do NOT provide explicit information on the three-dimensional geometry of a*

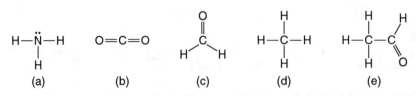

Figure 3-1 Lewis structures of various molecules. (a) ammonia, NH_3; (b) carbon dioxide, CO_2; (c) formaldehyde, H_2CO; (d) methane, CH_4; and (e) acetaldehyde, $H_3CCH{=}O$.

molecule. This was previously mentioned for H_2O. Another example is NH_3. The Lewis structure of NH_3 in Figure 3-1 implies that all of the atoms are in the same plane (the plane of the paper) and that the atoms are arranged in a T-shape. In actuality, neither of these is the case.

Valence shell electron pair repulsion (VSEPR) theory does a remarkable job of predicting the three-dimensional geometry about an atom. Although you probably covered VSEPR theory to some depth in general chemistry, I have found that many students fail to appreciate the power of such a simple model. Much of this is because general chemistry textbooks tend to provide a large table that summarizes the results of VSEPR theory, in which it correlates Lewis structure scenarios with the various molecular geometries (e.g., $AX_4 =$ "tetrahedral," $AX_2E_2 =$ "bent"). The problem is that those tables promote memorization; students tend to want to memorize the dozen or so results, rather than understand where they came from. This is particularly problematic, because those tables are typically not provided on exams.

VSEPR theory is quite straightforward and requires no memorization. The idea is that all electrons in a molecule (both bonding and nonbonding) exist as "groups" of electrons (e.g., a triple bond is one "group" of six electrons). Table 3-1 explicitly shows the four different groups of electrons found in organic molecules. Because groups of electrons are comprised of negatively charged particles, those groups repel each other and would prefer to be as far away from each other as possible. *The atoms that are at the ends of the covalent bonds must follow where the groups of electrons go, thus giving rise to the geometry about an atom.*

As an example, we can consider CO_2 (whose Lewis structure is provided in Figure 3-1), and ask what the geometry is about the central C atom. The electrons about the C appear as

Table 3-1 Electron groups in VSEPR theory

Electron Class	Number of e⁻ total	Number of "groups"
1 lone pair	2	1
1 single bond	2	1
1 double bond	4	1
1 triple bond	6	1

two double bonds. According to Table 3-1, that is two groups of electrons that repel each other. To get as far away from each other as possible, one group will be opposite to (180° apart from) the other group. In other words, the groups of electrons form a **linear** geometry about the C atom, or, the **electronic geometry** about the C atom is linear. Since an O atom is at the end of each double bond, the O atoms, too, are 180° apart. We therefore say that the O atoms are arranged linearly about the C atom, or, the **molecular geometry** is linear. It turns out, then, that the Lewis structure in Figure 3-1 is an accurate representation of the molecule's shape.

Next we can look at a molecule of formaldehyde, $H_2C{=}O$ (Figure 3-1). Surrounding the central C atom, we find two single bonds (attached to H atoms), and one double bond (attached to the O atom). According to Table 3-1, that is three total groups of electrons. For those groups to be as far away from each other as possible, they will end up pointing roughly to the corners of an equilateral triangle, being about 120° apart (give or take, depending on the effective sizes of the different groups). We therefore say that the electronic geometry is **trigonal planar.** Again, because there are atoms at the end of each group of electrons, we also say that the molecular geometry is trigonal planar, and the Lewis structure provided in Figure 3-1 accurately represents the molecule's geometry. Recognize that *trigonal planar* refers to the bond angle about the central C atom and also emphasizes the fact that the O and two H atoms are in the same plane as that C.

Methane, CH_4, is slightly more complex. In Figure 3-1, the Lewis structure makes it appear as if the entire molecule is in one plane, but that is not true. There are four groups of electrons surrounding the carbon atom. If all of them are constrained to the same plane, then they would be 90° apart, pointing to the corners of a square. If they are allowed to relax into a formation that is not all in one plane, they can be up to 109.5° apart. In such a case, each group points toward the corner of a tetrahedron, shown in Figure 3-2. The C atom is at the center of that tetrahedron, and an H atom is at each corner. Both the electronic and the molecular geometry are **tetrahedral.** Notice that because of a tetrahedron's symmetry, all four hydrogen atoms are equivalent. That is, they are all the exact same distance from the C atom, and every H—C—H angle is exactly the same at 109.5°.

Figure 3-2 Representation of CH_4 in its tetrahedral geometry.

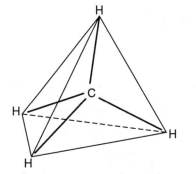

The aforementioned three molecules encompass the most common electronic geometries that appear in organic chemistry—linear, trigonal planar, and tetrahedral (the two other common ones you learn in general chemistry, trigonal bipyramid and octahedron, are not common in organic molecules). The lesson you should learn from the exercises we just went through is the *rigorous correlation among the number of electron groups, electronic geometry, and bond angle*. Specifically, if you know any one of the three, you know the other two. For example, if you know that the electronic geometry about an atom is trigonal planar, there must be a total of three electron groups, and any bond angle must be roughly 120°. Likewise, if you know that a bond angle is roughly 109.5°, then there must be a total of four electron groups about the central atom, and the electronic geometry must be tetrahedral. This idea is summarized in Table 3-2.

In each of the molecules we just examined with VSEPR theory, there are no lone pairs of electrons on the central atom. In other words, in those examples, there is an atom at the end of each group of electrons. That's why the geometry that describes the electron groups about the central atom (the electronic geometry) is the same as that which describes the positioning of the atoms about the central atom (the molecular geometry). However, *there exist several molecules that contain lone pairs of electrons, meaning that the electronic geometry cannot be the same as the molecular geometry*. In those circumstances, the electronic geometry is determined no differently than before (Table 3-1). The molecular geometry, however, must be considered further.

Ammonia, NH_3, is an example of a molecule that contains a lone pair of electrons on the central atom. The Lewis structure (Figure 3-1) shows that there are four groups of electrons on the N atom—three single bonds and one lone pair. Therefore, the electronic geometry must be tetrahedral (Table 3-2). There are H atoms at the ends of three of those four groups, as shown in Figure 3-3a. Consequently, the molecular geometry cannot be tetrahedral—that would require an atom at the end of the fourth group. The molecular geometry that is chosen for NH_3 is one that accurately describes the positioning of the atoms about the central N atom. Focusing only on the atoms (Figure 3-3b), it appears that the N atom is sitting on top of the three H atoms, and each H atom is located at the corner of an equilateral triangle. Therefore, the N atom and the three H atoms form a pyramid. The molecular geometry for NH_3 is **trigonal pyramid.**

The main difference between NH_3 and CH_4, as far as the application of VSEPR theory is concerned, is the molecular geometry. The rest is the same. Specifically, they both have the

Table 3-2 Correlations in VSEPR theory

# of Electron Groups	Electronic Geometry	Approx. Bond Angle
2	Linear	180°
3	Trigonal Planar	120°
4	Tetrahedral	109.5°

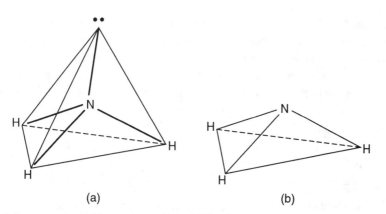

Figure 3-3 Representations of the geometry of NH₃. (a) The electronic geometry is tetrahedral. (b) The orientation of just the atoms is described as a trigonal pyramid.

same number of electron groups about the central atom. From Table 3-2, we see that they both have the same tetrahedral electronic geometry, and the angle between groups of electrons is roughly 109.5°. Furthermore, since the bond angle is the same as the angle between those electron groups, the H—C—H angle in CH_4 is about the same as the H—N—H angle in NH_3 (they are actually a few degrees different).

Each of the molecules we examined thus far in this section contain only one central atom. However, it may be that in a given molecule, there are two or more atoms about which we can define a geometry. A good example is acetaldehyde (Figure 3-1). One carbon atom has four single bonds, or four groups of electrons around it. Its electronic and molecular geometry are therefore tetrahedral. The other carbon atom, on the other hand, has three groups of electrons surrounding it—two single bonds and one double bond. Its electronic and molecular geometry are therefore trigonal planar.

We must point out that an atom can have an electronic geometry without a molecular geometry. Such is the case with either of the oxygen atoms on CO_2 (Figure 3-1). Each oxygen atom has three groups of electrons—one double bond and two lone pairs (the lone pairs are not shown in the Lewis structure). According to VSEPR theory, then, the electronic geometry about that atom must be trigonal planar, and the angle between any two groups must be roughly 120°. However, there is no bond angle that exists with that oxygen atom at the center. Therefore, there is no molecular geometry about the oxygen atom.

3.3 TETRAHEDRAL GEOMETRY AND THE DASH-WEDGE NOTATION

Tetrahedral geometry introduces three-dimensionality into organic molecules. You must therefore be able to draw and interpret their three-dimensional representations. This can often be difficult, given that the representations of three-dimensional structures are drawn in two dimensions. The *dash-wedge* notation was developed with the intention of making such a task straightforward. It consists of three aspects: (1) A straight line (——) represents a bond that is in the plane of the paper, such that the atoms at either end of the bond are both in the plane of the paper. (2) A wedge (——■) represents a bond that points toward you and comes out of the plane of the paper. In general, the atom at the thinner end of the wedge is in the plane of the paper, and the atom bonded at the thicker end is in front of the page. The idea is that if a tube (representing the bond) is pointed toward you, the end that is closer appears to be bigger. (3) A dashed line (⸺ or -----) represents a bond that is pointed away from you. Therefore, the central atom is in the plane of the paper, and the atom at the other end of the dashed line is behind the plane of the paper.

Using the dash-wedge notation, there are two common ways of representing a tetrahedral C atom. This is illustrated in Figure 3-4, which shows two representation of CH_4—a molecule whose geometry is a tetrahedron. It is important to realize that *both representations are of exactly the same molecule*, with the same 109.5° bond angles. The only difference is the vantage point from which you are viewing the molecule. Figure 3-4 emphasizes this, but there is no better way of understanding it than to actually *build a molecule of CH_4 using a molecular modeling kit!* Hold that molecule you build up to each of the two dash-wedge representations, and rotate the model until its orientation looks exactly like one representation. Then rotate the model again until it looks exactly like the other.

Being able to draw and to fully interpret the two different dash-wedge representations of a tetrahedral atom in Figure 3-4 is so important that we will go through some additional exercises that provide insight into the nature of a tetrahedral structure. The exercises specifically demonstrate three different ways to obtain a molecule of CH_4 and should help you understand the characteristics of a tetrahedron much better.

The first is to begin with a *fictitious* molecule of CH_4 that is square planar—that is, all five atoms are in the same plane (Figure 3-5). In Figure 3-5a, we view the planar molecule from the side, such that two H atoms and the C atom are in the plane of the paper, one H atom is pointing toward you, and the other is pointing away from you (pay close attention to the dash-wedge notation here). The H atoms on the left and right are then bent upward until the H—C—H bond angle is 109.5°. Likewise, the H atoms in front and in back are bent downward until that H—C—H bond angle is 109.5°. The result is the dash-wedge representations on the left in Figure 3-4, where the C and two Hs are in the plane of the paper, one H is in front, and one is behind.

Figure 3-4 Two different representations of CH_4 using the dash-wedge notation. Those representations imply viewing the molecule from different vantage points. The representation on the right is exactly how CH_4 would appear if viewed where indicated with the eyeball.

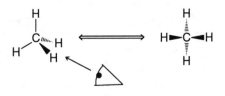

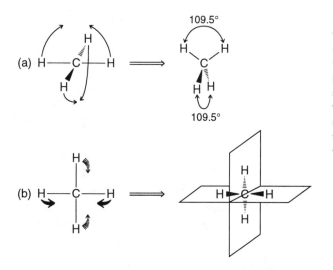

Figure 3-5 The tetrahedral geometry of CH_4 can be obtained beginning with a fictitious molecule of CH_4 that is square planar. (a) The planar molecule is viewed from the side. (b) The planar molecule is viewed from the top. In this representation, we can most readily see the two perpendicular planes in a tetrahedral structure.

In Figure 3-5b, we view the original planar molecule from the top and repeat the process of shifting the H atoms. In this case, a pair of opposite H atoms is bent toward you, resulting in wedges, while the other pair of opposite H atoms is bent down away from you, resulting in dashed lines. The result is the dash-wedge representation on the right in Figure 3-4.

One of the main reasons for going through that exercise is to realize that in a tetrahedral molecule like methane, *there are two perpendicular planes* that are defined. One plane is defined by the C and a pair of Hs (three points are needed to define a plane). The second plane is defined by the C and the other pair of Hs. The fact that those planes are perpendicular to each other is most easily seen in Figure 3-5b. When viewing the tetrahedral structure from that vantage point, it appears that the line connecting one pair of Hs is perpendicular to the line connecting the other pair. The plane that contains the C and the first pair of Hs must therefore be perpendicular to the plane that contains the C and the second pair of Hs.

In light of these perpendicular planes, you can envision constructing a molecule of CH_4 using two V-shaped pieces, each of whose angle is 109.5° (Figure 3-6). Specifically, orient those pieces so that the plane of one is perpendicular to the plane of the other, and the opening of each V is pointed in opposite directions—e.g., toward each other. Bring the two

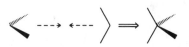

Figure 3-6 A tetrahedral atom can be constructed from two V-shaped pieces, each with an angle of 109.5°. One V-shaped piece is oriented perpendicular to the second before they are brought together. In the resulting structure, the point at which the Vs are fused together represents the central atom.

Figure 3-7 Construction of a tetrahedral molecule from a perfect cube.

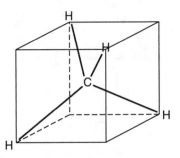

pieces together and fuse them at the vertices of the Vs. The fusion at the vertices becomes the tetrahedral C. (It is worth noting that one brand of molecular modeling kits takes advantage of this. Two V-shaped pieces, snapped together in exactly the way described here, yields a single tetrahedral atom.)

Another way of seeing the perpendicular planes within a molecule of CH_4 comes from a second way of constructing the tetrahedral structure. Begin with a perfect cube, where the C atom is exactly in the center (Figure 3-7). The cube is comprised of a total of six faces (top, bottom, left, right, back, and front), and eight corners (four on top and four on the bottom). The four H atoms are placed at four of those eight corners. Two H atoms are placed at corners on the top face, opposite each other. The other two H atoms are placed at corners on the bottom face, also opposite each other, such that neither of the second two H atoms is directly below any H atom from the first pair. Viewed from the top, the molecule would then appear as the representation on the right in Figure 3-5b.

The third way to construct CH_4 demonstrates a different aspect of the molecule's symmetry. Realize from Figure 3-2 that a tetrahedron itself has four faces, and each face is an equilateral triangle (all three sides and all three angles are identical). The C atom is at the center of the tetrahedron. Therefore, if we ignore the topmost H atom and focus only on the C atom and the bottom three H atoms, we appear to have a pyramidal structure (Figure 3-8). The C atom sits on top of the bottom three H atoms. The molecule of CH_4 can therefore be viewed as a pyramid constructed of one C atom and three H atoms, with a fourth H atom placed directly on top of that pyramid.

Figure 3-8 (a) Construction of CH_4 by placing an H atom on top of a pyramid containing a C atom and three H atoms. (b) View of CH_4 from the top, explicitly showing the H atoms at the corners of an equilateral triangle. Therefore, if the CH_4 molecule is rotated 120° about the topmost C—H bond (indicated by the curved arrow), the molecule looks no different.

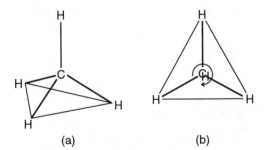

(a)   (b)

There are two lessons to learn from the last exercise. One is that *the CH_4 molecule is perfectly symmetric.* That is, we could have chosen any set of three H atoms to use as the pyramid's base, because all four faces of a tetrahedron are identical. The second and, perhaps more important, lesson is to understand that *a molecule of CH_4 has what is called "three-fold rotational symmetry."* What that means is that if we rotate the molecule 360° about a certain axis, it appears to be the same at three different angles of rotation—120°, 240°, and 360°. This rotational symmetry can be seen in Figure 3-8b, where we view the molecule from the top. The axis about which we would rotate would be a carbon-hydrogen bond, as indicated in the figure.

3.3a Pitfalls with the Dash-Wedge Notation

One of the most common mistakes pertaining to the dash-wedge notation is to take it for granted. A student might intend to draw a tetrahedral atom using only straight lines and omitting the dashes and wedges, such as that shown in Figure 3-9a. The intention might be to imply, simply by the positioning around the central atom, that certain atoms or groups of atoms are in front of the plane of the paper, while others are behind. However, if the dash-wedge notation is not explicitly shown, then the information provided to anyone interpreting the structure is no different than that provided by the Lewis structure—that is, only the connectivity (which atoms are connected to which, and by what type of bonds) is actually provided.

Conversely, when interpreting a structure that does not contain dashes and wedges explicitly, a student might infer that, because an atom or group of atoms appears at a certain location around the central atom, then it must be in front of the plane of the paper. They might infer that another atom is behind. As before, unless a tetrahedral atom is drawn explicitly with the dashes and wedges, those inferences are not valid.

Another common mistake with the dash-wedge notation arises despite good intentions by the student. That is, the student includes the dashes and wedges, but does so incorrectly. Figure 3-9b illustrates the most common incorrect usage of the dash-wedge notation. If we recall (see Figure 3-6) that the tetrahedral C atom is the result of fusing two Vs together, one whose opening is in the opposite direction as that of the second, then we should see right away that the representation in Figure 3-9b is invalid. In Figure 3-9b, the Vs are not pointing in opposite directions. The V that is in the plane of the paper opens downward.

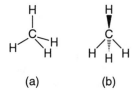

Figure 3-9 Common mistakes in trying to depict a tetrahedral C in three dimensions. (a) Dash-wedge notation not used. (b) Dash-wedge notation used incorrectly. To correct the mistake, both of the bonds not in the plane of the paper should be pointing upward.

(a)   (b)

The other two bonds, comprising the second V (perpendicular to the first), *should* open in the upward direction. In other words, both of the bonds that are not in the plane of the paper should be pointing upward. In Figure 3-9b, one points upward and the other points downward, resulting in a representation that does not accurately depict the true tetrahedral structure. Because of this, such a structure is, once again, no more useful than a simple Lewis structure.

Why are the previous mistakes important to avoid? The answer primarily has to do with stereochemistry—something that we will discuss in Chapter 4. Stereochemistry requires accurate three-dimensional representations of tetrahedral atoms. *The bottom line is that if you make mistakes such as those in Figure 3-9, your answer as to the stereochemistry of a molecule will be wrong 50 percent of the time!*

3.4 POLARITY AND BOND DIPOLES

We learned earlier that atoms can be bonded together through the sharing of a pair of electrons—that is, a covalent bond. If a bond is perfectly covalent, the pair of electrons is shared equally. Such is the case in a molecule of H_2. In the H—H bond, the pair of electrons is shared between identical atoms. Therefore, there is no reason for the electrons to prefer one H atom over the other.

The pair of electrons in the H—H bond is shared equally because the H atoms have exactly the same electronegativity. **Electronegativity** can be defined as an atom's *pull of electrons toward itself, when that atom is part of a covalent bond*. Recall that electronegativity is one of the periodic table trends that you learned in general chemistry; it *increases up the periodic table, and also from left to right across the periodic table*. No two atoms have exactly the same electronegativity. Therefore, a *covalent bond between atoms that are not identical must not share the pair of electrons equally*.

A molecule of H—F, for example, must not have equal sharing of the electrons between the H and F atoms. According to the periodic table trend, the F atom has a much higher electronegativity than H, and, as a result, has a much stronger pull of the shared electrons toward itself. If that's true, then F has more than its fair share of the pair of electrons, and the F atom bears a partial negative charge (δ^-), as shown in Figure 3-10. Since the overall molecule is neutral, there must be a balance of positive charge on the H atom (δ^+). Such *a*

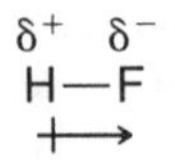

Figure 3-10 The bond dipole of the H—F bond. (top) Represented by partial positive and negative charges on the H and F atoms, respectively. (bottom) Same depiction of the bond dipole, using the arrow notation.

separation of charge, caused by the unequal sharing of electrons in a covalent bond, is called a **bond dipole moment,** or **bond dipole** for short.

To get a better feel for a given bond dipole, an arrow is drawn that explicitly shows the direction in which the bonding electrons are being pulled (Figure 3-10). The head of the arrow is therefore analogous to the δ^-, and the tail of the arrow is analogous to the δ^+. To remind you of this, a short line is drawn through the tail of the arrow to resemble a "+".

3.5 NET MOLECULAR DIPOLE MOMENT

If there are bond dipoles present in a molecule, each of which contributes a δ^+ and a δ^-, then it is possible for one end of the overall molecule to bear a δ^+, while the other end bears an equal and opposite δ^-. In other words, there can be a **net molecular dipole moment,** making the entire molecule **polar;** a trivial example is the case of HF mentioned previously, given that its only bond is a polar covalent bond. On the other hand, those partial charges can also add up in a certain way so as to completely cancel one another—that is, the molecule may be **nonpolar.**

The net molecular dipole is determined by adding up all of the bond dipoles within the molecule. For those of you who are comfortable with vectors, we treat each bond dipole as a vector, and add them up "vectorially" to determine the overall molecular dipole moment. For those of you who are not, think about a tug-of-war. Each bond represents a rope tying two atoms together, and each bond dipole represents a tug-of-war between those two atoms; the bond dipole arrow is drawn in the direction of the atom that is pulling stronger. In a molecule with several bonds, each with a bond dipole of its own, think about several ropes and several tugs-of-war going on at the same time. If all of those imaginary tugs-of-war would result in the overall molecule being pulled in one direction, then the molecule is polar. *The direction of the net molecular dipole is the same as the direction in which the molecule would move if all of those imaginary tugs-of-war were going on.*

To help you better understand this idea, let's determine the net molecular dipole for each molecule in Figure 3-11. The first is CO_2 (Figure 3-11a). From VSEPR theory, we know that the geometry is linear about the C atom and that the O atoms are on opposite sides. There are two bond dipoles in this molecule—one for each C=O bond. One is pointed from the C toward one O atom, and the other is pointed from the C toward the second O atom. Treating each bond dipole as a tug-of-war between an O and the C, it appears that the C atom is being pulled equally in opposite directions. The result of our imaginary tugs-of-war is that the molecule does not move, and we associate this with the molecule being nonpolar.

Next, we have F—C≡N (Figure 3-11b), which is also a linear molecule. Both the F and the N atoms are more electronegative than C, such that the FC and CN bond dipoles originate from the C atom and point to the F and N, respectively. In terms of our imaginary tugs-of-war, the F atom is pulling the C atom to the left, and the N atom is pulling the C atom to the

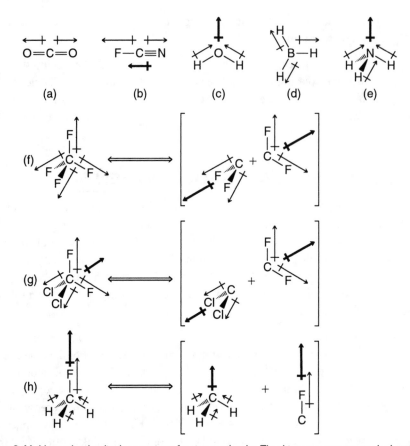

Figure 3-11 Net molecular dipole moments of various molecules. The thin arrows represent the bond dipoles, and the thick arrows represent the net molecular dipoles. Absence of a thick arrow indicates no net molecular dipole. In (f) and (g), the tetrahedral C atom is split up into its two V-shaped pieces (inside the brackets), each of which has a net dipole—in (f) they cancel, but in (g) one is stronger than the other. In (h), the tetrahedral C is split up into a trigonal pyramid and a C—F bond, each of which has a net dipole—they both point in the same direction and therefore add together.

right. Since the F atom is more electronegative than N, it is pulling the C atom harder, and the entire molecule is pulled to the left. Therefore, we say that the molecule is polar, and the net molecular dipole points to the left.

From the discussion at this chapter's beginning, we already know that H_2O (Figure 3-11c) is a polar molecule. This is because it is a bent molecule, composed of two bond dipoles. Both originate from an H atom and point toward the O. Rather than think about it in terms of tug-of-war, consider it a "push-of-war," where both H atoms are pushing on the O atom. One H is pushing the O atom up and to the right, and the other is pushing it up and to the left. The result is the movement of the entire molecule directly upward, which is the same as the direction of the net molecular dipole.

BH_3 (Figure 3-11d) is a trigonal planar molecule, with the H atoms at the corners of an equilateral triangle, and the B atom at the center. There are three B—H bond dipoles to consider. It turns out that hydrogen has a greater electronegativity than boron (this is difficult to tell by the periodic table trend due to the placement of H in the periodic table), so those bond dipoles originate from the B atom and point toward the different H atoms. Now consider those bond dipoles as tugs-of-war, where each H atom is tied to the B and pulls it toward the corner of the equilateral triangle. Because of the symmetry of an equilateral triangle, the end result would be no movement of the molecule whatsoever, and we say that there is no net molecular dipole.

Next, we examine NH_3 (Figure 3-11e), which is a pyramidal molecule. The H atoms form an equilateral triangle, and the N atom is sitting on top of them, at the triangle's center. Each of the N—H bond dipoles points from an H atom to the N. As with H_2O, think of these as pushes-of-war, where each H atom is pushing on the N atom. That is, each H atom simultaneously pushes inward and upward. The horizontal (inward) forces cancel out because each H atom pushes from the corner of an equilateral triangle. As a result, there would be no motion in the horizontal direction. What is left is the upward force from each H atom. The overall resulting motion of the entire molecule would therefore be straight upward, the same as the direction of the net molecular dipole.

In the final examples, we will consider three molecules with a tetrahedral central C atom. The first is CF_4 (Figure 3-11f), where each of the F atoms is at the corner of a tetrahedron, and the C atom is at the center. Because F is more electronegative than C, each F atom can be thought of as pulling on the C atom. Would there be any resulting movement of the molecules? The answer is no, because of the symmetry of a tetrahedron. We can see this better if we split CF_4 into two identical V-shaped portions (similar to Figure 3-6), where C is at the vertex, and an F is at each end. For each V-shaped piece, then, the net dipole is pointing toward the opening of the V. Because CF_4 is composed of two identical Vs, pointing in opposite directions, the result is no net dipole.

Next let's examine CF_2Cl_2 (Figure 3-11g). Again, we can split this molecule up into two V-shaped pieces—a CF_2 and a CCl_2. As before, the net dipole from each V points in the opposite direction. However, because the bond dipole of a C—F bond is stronger than that of a C—Cl bond, the V formed by the CF_2 pulls stronger than that formed by the CCl_2. As a result, the net molecular dipole points in the direction of the opening of the V formed by the CF_2.

Finally, let's consider CH_3F (Figure 3-11h). We can separate this molecule into an F atom and a pyramidal structure composed of CH_3 (with C on top of the three Hs). Each of the C—H bond dipoles points toward the C, so let's think of each H pushing on the C. The resulting movement of the CH_3 group would be directly upward, similar to what we saw with NH_3. If we then tack on the F, we simply add on the bond dipole of the C—F bond, which points toward the F. The pull of the F atom is in the same direction as the movement of the CH_3 group, so that the overall movement of the CH_3F molecule would be directly upward. Therefore, the net molecular dipole is along the C—F bond, pointing in the direction of the F atom.

49

3.6 HYBRIDIZATION

Hybridization is closely tied with molecular orbital (MO) theory (Section 3.7). Together, those two models are complementary to Lewis structures and VSEPR theory. Whereas MO theory provides information about bonding within a molecule (as do Lewis structures), hybridization (like VSEPR theory) provides information about molecular geometry. MO theory and the concept of hybridization are more powerful models than Lewis structures and VSEPR theory, in that the former can be employed to make accurate predictions about the behavior of molecules that the latter cannot.

There are three types of hybridization that we must be concerned with: sp, sp^2, and sp^3. All three involve the mixing together of s and p atomic orbitals (AOs) from the valence shell of an atom (for C and other second row elements, this will be the 2s and 2p orbitals) to form what are called **hybridized atomic orbitals.** Think about it as each of those pure s and p valence shell orbitals being *transformed* into a hybridized orbital. As we will see later, the motivation for forming hybridized AOs is that they are better for bonding than the unhybridized (s and p) orbitals.

Because s and p AOs are used for hybridization, we should first examine some of the AO characteristics—in particular, the shape, phase, and orientation (Figure 3-12). All s orbitals are spheres, the center of which is the location of the atom's nucleus. There is really no orientation to consider, because spheres look exactly the same no matter how they are rotated. They come

Figure 3-12 (a) An s orbital is spherical, with the nucleus (small dot) at the center. The phase is either unshaded (left) or shaded (right). (b) A p_x orbital, with a spherical lobe on either side of the nucleus. The phase of one lobe is opposite the other. Either the left lobe is unshaded while the right lobe is shaded (left), or the left lobe is shaded while the right lobe is unshaded (right). (c) Two schemes for the phase of a p_y orbital. (d) Two schemes for the phase of a p_z orbital. One lobe is in front of the other, with the nucleus (small dashed circle) in-between.

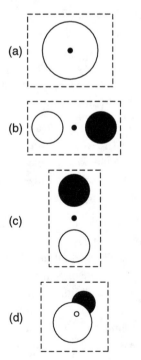

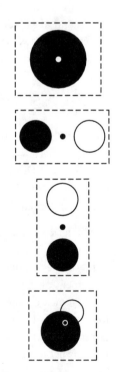

in one of two **phases,** which are viewed as opposite to one another. The phase of an orbital does have physical meaning, but there is no need to discuss it here. We will simply call those phases "shaded" or "unshaded." Both phases—the shaded and unshaded—are identical in every way, except for how they interact with another orbital, as we will see shortly.

The shape of a p orbital (Figure 3-12b, c, or d) is a "dumbbell" with two "lobes." Each lobe is spherical* —it is three-dimensional. The atom's nucleus is located between those lobes. A p orbital exists in one of three orientations. The lobes can be aligned along the x-axis (Figure 3-12b, one lobe on the left of the nucleus and one on the right), the y-axis (Figure 3-12c, one lobe above the nucleus and one below), or the z-axis (Figure 3-12d, one lobe in front of the nucleus and one behind). Those three different orbitals are called p_x, p_y, and p_z, respectively. As far as phase is concerned, one lobe is shaded while the other is unshaded. Just as with s orbitals, there are two ways to do the shading for each p orbital. For example, in the p_x orbital, the left lobe could be shaded, while the right is unshaded. Alternatively, the left lobe could be unshaded, while the right is shaded.

As its name suggests, **sp hybridization** involves *the mixing together of the s and a single p orbital from the valence shell.* Let's arbitrarily choose the p_x orbital, leaving the p_y and p_z alone—that is, leave the p_y and p_z orbitals *unhybridized.* There are two ways in which the s and p_x orbitals can mix together (Figure 3-13) and both occur simultaneously, resulting in

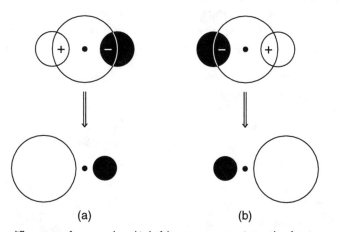

(a)   (b)

Figure 3-13 Two different ways for an s and p orbital of the same atom to mix together, forming two sp-hybridized AOs. (a) Top: The phases of the s and p orbitals are the same on the left of the nucleus, and opposite on the right of the nucleus. There is therefore "good" overlap on the left, indicated by "+," and "bad" overlap on the right, indicated by "–." Bottom: The resulting sp-hybridized orbital is built up on the left and diminished on the right. (b) Top: The phases of the s and p orbitals are the same on the right of the nucleus, and opposite on the left of the nucleus. The "good" overlap is therefore on the right, and the "bad" overlap is on the left. Bottom: The resulting sp-hybridized orbital is built up on the right and diminished on the left

A more accurate description of a p orbital is a "distorted hourglass," where the lobes are close to, but not exactly, spherical. However, taking the lobes to be spherical does not qualitatively change any results presented in this book.

two new **sp-hybridized orbitals.** In Figure 3-13a, the phase of the s orbital is the same as the phase of the left lobe of the p orbital and is opposite in phase to the lobe on the right. In Figure 3-13b, the reverse is true.

If we mix the two orbitals together as in Figure 3-13a, there is "good" overlap between the s and p orbitals to the left of the nucleus—that is, the phase of the orbitals is the same in that region. On the other hand, there is "bad" overlap to the right of the nucleus, because the orbitals are of opposite phase in that region. The hybridized AO that is formed is significantly built up on the left, as a result of the "good" overlap, and is diminished on the right, as a result of the "bad" overlap. Keep in mind that the sp-hybridized orbital is three-dimensional, even though it is depicted in two dimensions. Think of it as a large sphere on the left and a small sphere on the right.*

The second sp-hybridized orbital is derived in exactly the same way, by the mixing of the orbitals in the manner shown in Figure 3-13b. In this case, the "good" overlap is on the right, and the "bad" overlap is on the left. The resulting hybridized AO is therefore built up on the right, whereas it is diminished on the left.

To envision the generation of sp-hybridized orbitals, imagine both the s and the p orbitals made from clay. Suppose the s orbital is made from red clay, and the p orbital is made from blue clay. To make the sp-hybridized orbitals, we would mix the red clay and the blue clay together to make purple clay, and we pull them apart to make new sp-hybridized orbitals. Clearly, then, these hybridized orbitals are distinct from s and p orbitals, but they have characteristics of each. Moreover, given that there are two orbitals' worth of clay mixed together, two orbitals are formed when the clay is pulled apart. We can view this as a *conservation of orbitals.*

In showing the explicit derivation of the sp-hybridized orbitals, we arbitrarily used the s orbital and the p_x orbital from the valence shell, which generated two hybridized orbitals aligned along the x-axis. The p_y and the p_z orbitals were left alone and remained unhybridized. As a result, the valence shell would be comprised of the two sp-hybridized orbitals, along with the p_y and the p_z orbitals (Figure 3-14a).

What would have happened if we used either the p_y or the p_z orbitals for hybridization, instead of the p_x? To answer this question, we must remember that the only difference between the three different p orbitals is their orientation in space. Therefore, had we used either the p_y or p_z orbitals, the resulting orbitals should look exactly the same but oriented differently in space. For example, if we used the p_y orbital for hybridization, then the two hybridized orbitals would be aligned along the y-axis, and the p_x and p_z orbitals would remain unhybridized (Figure 3-14b). If, on the other hand, we used the p_z orbital for hybridization, then the two hybridized orbitals would be aligned along the z-axis, and the p_x and p_y orbitals would remain unhybridized (Figure 3-14c).

As before, the lobes are not perfectly spherical, but this does not qualitatively change any of our results.

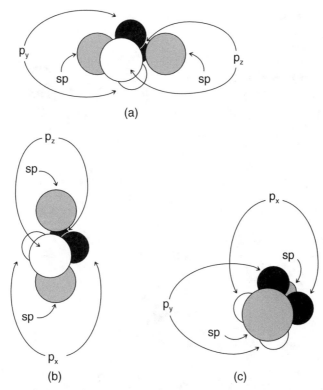

Figure 3-14 Atoms that have undergone sp hybridization. For simplicity, only the large lobe of each hybridized orbital (gray) is shown. (a) The p_x orbital was used for hybridization, leaving the p_y and p_z orbitals unhybridized. (b) The p_y orbital was used for hybridization. (c) The p_z orbital was used for hybridization.

The lesson to learn from Figure 3-14 is that no matter which p orbital is used for sp-hybridization, the *sp-hybridized orbitals must be aligned perpendicular to the two unhybridized p orbitals*. This is true no matter which way a pair of sp-hybridized orbitals is oriented.

In **sp² hybridization,** *there are three atomic orbitals in the valence shell that mix together— the s orbital and two of the three p orbitals*. The third p orbital is left unhybridized. Because sp² hybridization involves mixing together three total orbitals, the result is *three new hybridized AOs*, called **sp²-hybridized orbitals.** Again, think of the clay analogy, where we would mix together an s orbital made out of red clay and two p orbitals made out of blue clay. The result would again be purple clay (this time, a bit more blue than red), which we can pull apart and make into three sp²-hybridized orbitals (recall the conservation of orbitals).

Unlike sp hybridization, it is not easy to show the explicit derivation of the shape of an sp² orbital using the notion of "good" and "bad" overlap. However, all three sp² orbitals have exactly the same shape as one another and are very similar in shape to sp-hybridized

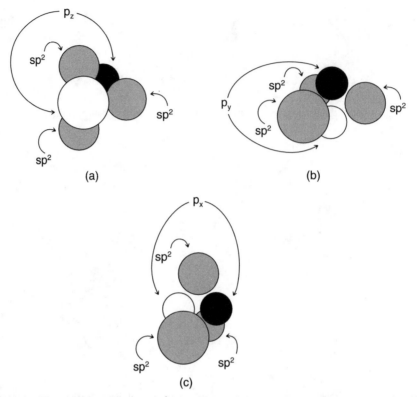

Figure 3-15 Atoms that have undergone sp^2 hybridization. (a) The p_x and p_y orbitals were used for hybridization, confining the hybridized orbitals in the x–y plane and leaving the p_z orbital unhybridized. (b) The p_x and p_z orbitals were used for hybridization. (c) The p_y and p_z orbitals were used for hybridization.

orbitals. That is, each sp^2 orbital is comprised of two spherical lobes—one large and one small. Therefore, sp^2 orbitals are directional, and, as it turns out, they point to the corners of an equilateral triangle, 120° apart (Figure 3-15).

Similar to sp hybridization, it is arbitrary as to which two p orbitals are used for hybridization with the s orbital. Just as with sp hybridization, however, the different choices simply result in different orientations of the resulting orbitals. For example, if the p_x and p_y orbitals are used for hybridization, then the three resulting sp^2-hybridized orbitals must lie in the x–y plane, and the p_z orbital is left unhybridized (Figure 3-15a). If, on the other hand, the p_x and p_z orbitals were used for hybridization, the resulting sp^2-hybridized orbitals must lie in the x–z plane, and the p_y orbital is left unhybridized (Figure 3-15b). Similarly, if the p_y and p_z orbitals are used for hybridization, the sp^2-hybridized orbitals must lie in the y–z plane, leaving the p_x orbital unhybridized (Figure 3-15c).

In examining Figure 3-15, it is clear that the difference is only in the orientation of the species. The lesson here is simply that whatever the choice of p orbitals, *the plane*

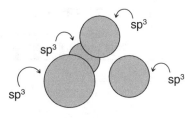

Figure 3-16 The four sp³-hybridized orbitals, pointing toward the corners of a tetrahedron.

defined by the three sp²-hybridized orbitals must be perpendicular to the single unhybridized orbital.

In **sp³ hybridization,** all four valence shell orbitals are mixed together—the s orbital and all three p orbitals. *The result is four new sp³-hybridized orbitals,* leaving no unhybridized orbitals in the valence shell. All four sp³-hybridized orbitals are qualitatively the same as sp and sp² orbitals; they are directional, with one large spherical lobe and one small one. The four larger lobes point to the corners of a tetrahedron, 109.5° apart (Figure 3-16).

In sp³ hybridization we need not worry about how the orientation of the resulting orbitals depends upon the choice of p orbitals involved in hybridization. This is because all three of the valence shell p orbitals are used for sp³ hybridization—there is no choice to begin with.

Recall that the motivation for hybridization is for an atom to have AOs that are better for bonding. Although we will not fully understand why that is so until after we discuss MO theory (which deals specifically with bonding), we can begin to understand it from the *directionality* of hybridized orbitals. Unhybridized s and p orbitals are symmetric and therefore nondirectional. On the other hand, hybridized orbitals are directional. For example, the sp-hybridized orbitals in Figure 3-14 "point" in opposite directions—that is, 180° apart. Similarly, sp²-hybridized orbitals are directional, in that they "point" 120° apart, and sp³-hybridized orbitals "point" 109.5° apart.

Because we anticipate hybridized orbitals to be used for bonding, we should expect that the directionality of those hybridized orbitals should lead to specific bond angles. For example, given that the two sp-hybridized orbitals are 180° apart, we ought to expect an sp-hybridized atom (an atom whose valence s and p orbitals have undergone sp hybridization) to have a linear electronic geometry. We should also expect that the 120° angle between sp²-hybridized orbitals in an sp²-hybridized atom should give rise to a trigonal planar electronic geometry and that the 109.5° angle between sp³-hybridized orbitals in an sp³-hybridized atom should give rise to a tetrahedral geometry.

There appears to be a clear connection between hybridization and VSEPR theory, as summarized in Table 3-3. It should therefore be straightforward to identify the type of hybridization of any atom in a molecule. The number of electron "groups"

Table 3-3 Relationship between hybridization and VSEPR theory

Hybridization	Bond Angle	Electronic Geometry
sp	180°	Linear
sp^2	120°	Trigonal Planar
sp^3	109.5°	Tetrahedral

an atom has dictates its electronic geometry, as governed by VSEPR theory. From the electronic geometry, the type of hybridization is readily identified, as shown in Table 3-3.

3.7 MOLECULAR ORBITAL (MO) THEORY

The main idea behind MO theory is that all bonding electrons are confined to orbitals called **molecular orbitals.** As with any orbitals, each MO can house up to two electrons—one that is spin up, and one that is spin down. Therefore, a single bond, which contains two total electrons, is comprised of one MO. A double bond, which contains a total of four electrons, is comprised of two MOs. And a triple bond is comprised of three MOs.

The formation of MOs is similar to the formation of hybridized atomic orbitals (AOs), in that they both involve mixing together orbitals to form new orbitals. The greatest difference between the two processes is that hybridization involves mixing AOs (s and p orbitals) from the same atom, whereas MOs involve mixing AOs (either hybridized or unhybridized) from *different* atoms. When two atoms are brought close together (a bond length apart), AOs from the different atoms overlap somewhat in space. Depending on the phases of those orbitals, that overlap could be "good" or "bad." *If the overlap is "good," a **bonding MO** is formed, and if the overlap is "bad," an **antibonding MO** is formed.*

For the purposes of this book, we will focus on only bonding MOs. The details of the antibonding MOs will be left to a traditional textbook, largely because they are rarely invoked. What we will specifically focus on are the shapes of the various types of bonding MOs that can be formed. As we will see in an application at the end of this chapter, it is the shapes of MOs that dictate whether or not atoms can rotate about a bond.

The first type of bond we examine is that formed when two s orbitals overlap—one from each atom involved in the bond. Figure 3-17a shows the region between the atoms in which there is "good" overlap. The result is an orbital that has been built up between the two atoms and looks like an egg (remember, it is three-dimensional) that encompasses both nuclei. The MO has most of its volume between the two atoms, since that allows a pair of electrons occupying that MO to spend most of the time there—something that we would expect if that pair of electrons is to be shared.

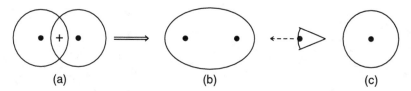

Figure 3-17 (a) "Good" overlap of an s orbital from each of two atoms, indicated by "+". (b) The result is a MO that looks like an egg. (c) When viewed down the bonding axis, the egg's outline is a circle. This MO is therefore designated a σ MO.

We must understand what bonding MOs look like when viewed down the bonding axis—the line connecting the two atoms. In this case, simply imagine what an egg would look like if viewed from one end—its outline would be a circle (Figure 3-17b–c). Notice that the outline of an s AO (a sphere) is also a circle. The MO formed from the overlap of two s AOs is therefore designated a σ (sigma, the Greek letter for s) MO, or an MO of σ symmetry. It is often simply called a σ bond.

A second type of MO is formed from the overlap of two hybridized AOs, one from each atom. There are six different combinations of overlapping hybridized AOs—sp with sp; sp with sp^2; sp with sp^3; sp^2 with sp^2; sp^2 with sp^3; and sp^3 with sp^3. However, we are only concerned with the general shape of the resulting MO. Because all three types of hybridized orbitals are qualitatively the same (i.e., they have a large spherical lobe on one end and a small spherical lobe on the other), all six combinations of overlap produce MOs that essentially look the same.

Figure 3-18a shows the overlap of two hybridized AOs, where there is a large amount of "good" overlap between the two atoms. The resulting MO is, once again, built up in that region (Figure 3-18b). That MO's shape can be thought of as a large egg between the two atoms, with much smaller spheres on either side. If viewed down the bonding axis (Figure 3-18c), we would see a small sphere in front of the egg. The outline of each would be a circle. Therefore, for the same reason as with the previous MO, this, too, is designated as a σ bonding MO.

A third type of MO is formed by the overlap between a hybridized AO from one atom, and an s orbital from the second (Figure 3-19a). The resulting MO looks like a large egg next to

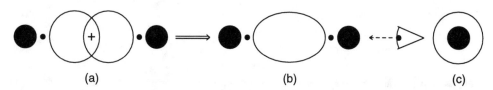

Figure 3-18 (a) "Good" overlap between two hybridized AOs, one from each of two atoms. (b) The resulting MO looks like an egg in-between two small spheres. (c) View of the MO down the bonding axis. Because the outline is a circle, it is a σ MO.

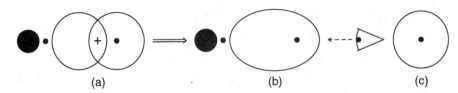

Figure 3-19 (a) "Good" overlap between a hybridized AO from one atom and an s orbital from the second. (b) The resulting MO looks like an egg next to a small sphere. (c) When viewed down the bonding axis, the outline is a circle, making this a σ MO.

a small sphere (Figure 3-19b). When viewed down the bonding axis, the outline is, once again, a circle (Figure 3-19c). This MO is therefore a σ MO.

The last type of MO we will examine is formed from the overlap between two unhybridized p orbitals (Figure 3-20). There are several ways of orienting these orbitals relative to one another, but the important one is that in which they are side by side and in which their phases match up (Figure 3-20a). In that orientation, there is "good" overlap between the spherical lobes both above and below the two nuclei. The result is *a single MO that looks like two eggs* (Figure 3-20b)—one above the two nuclei and one below. If viewed down the bonding axis, this MO looks different from the previous three. Each egg has an outline of a circle; there is one circle above the nuclei and one circle below. Such an outline resembles the outline of a p orbital. The symmetry of this type of MO is therefore designated as π (pi, the Greek letter for p). It is a π MO, giving rise to what is called a π bond.

One of the most common mistakes with π orbitals is interpreting each one as being comprised of two bonds. It is easy to see why this mistake is so often made—that bond is comprised of two distinct lobes, one on either side of the bonding axis. However, it is important that you see this as only one bond. It is no different from a p atomic orbital (the dumbbell shape) being one orbital that happens to be comprised of two distinct lobes.

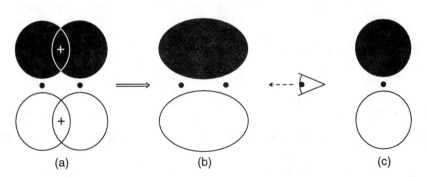

Figure 3-20 (a) Overlap between unhybridized p orbitals, demonstrating "good" overlap both above and below the two nuclei. (b) The resulting MO looks like two eggs, one above the two nuclei, and one below. (c) The MO viewed down the bonding axis. Each egg has an outline that is roughly circular, which overall resembles the outline of a p AO. This MO is therefore of π symmetry.

3.8 MOLECULAR ORBITAL PICTURES OF H_2, ETHANE, ETHENE, ETHYNE, AND FORMALDEHYDE

When we introduced MO theory, we indicated that covalent bonds are actually pairs of electrons occupying MOs. The four molecules H_2, ethane (H_3C—CH_3), ethene (H_2C=CH_2), and ethyne (HC≡CH) are classic molecules to examine in MO theory, because together they contain all four types of MOs we discussed in the previous section. Formaldehyde is also examined in this section because it provides an example of a molecule containing lone pairs of electrons.

The single bond in H_2, for example, is the result of a pair of electrons occupying the σ bond, constructed from an s AO donated by each H atom (Figure 3-17). Those AOs are donated because in the isolated H atom, the electron occupies the 1s orbital. When two hydrogen atoms are close together, those are the important orbitals that overlap in space.

To examine the MOs in H_3C—CH_3, we first use VSEPR theory to identify the type of hybridization the carbon atoms have undergone. The geometry about each C atom is tetrahedral; from Table 3-3 we can immediately see that each C is sp^3 hybridized, containing four hybridized orbitals in a tetrahedral arrangement (Figure 3-16). The same is true for the second C atom. In order to form a bond between the two C atoms (as we know there must be one), one hybridized orbital from each C atom must point to the other C atom. That sets up "good" overlap between the two sp^3 orbitals, resulting in the σ bond similar to that shown in Figure 3-18. That leaves six hybridized orbitals (three on each C atom) still available for bonding. From the Lewis structure of H_3C—CH_3, we know there must be a total of six C—H single bonds. Each of those bonds is attained by the overlap of one hybridized orbital from C and the s orbital from an H atom. We can now see that those C—H single bonds are, in fact, σ bonds similar to those shown in Figure 3-19.

There is a lot going on in Figure 3-21. If you do not fully understand it, carefully read the last paragraph one more time. Specifically focus on how it is we know that the C—C single bond is a σ bond between two sp^3 orbitals. Also, understand why it is that each C—H single bond is a σ bond between an sp^3 orbital on C and an s orbital on H.

In another example, we derive the type of bonding in H_2C=CH_2. We again begin by employing VSEPR theory to derive that each C atom is trigonal planar and therefore sp^2 hybridized (Table 3-3). Therefore, each C atom has three hybridized orbitals pointing to the corners of an equilateral triangle, and one unhybridized p orbital is perpendicular to that plane (Figure 3-15). To ensure that the C atoms are bonded together by at least one bond, a hybridized orbital from each C atom is pointed toward the other C atom, resulting in a σ bond of the type in Figure 3-18. As before, each of the four C—H single bonds is a σ bond of the type in Figure 3-19.

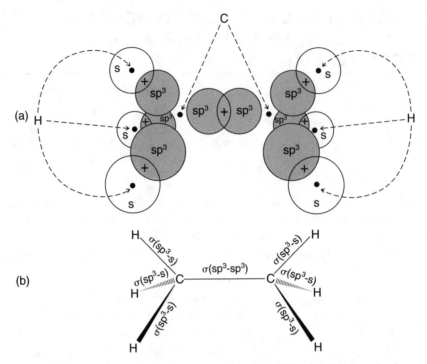

Figure 3-21 H_3CCH_3. (a) Overlap between atomic orbitals to form the bonds within the molecule. Hybridized orbitals belonging to the C atoms are gray. "Good" overlap is indicated by "+". (b) Dash-wedge notation. The C—C bond is a σ bond from the overlap between an sp^3 orbital on each C atom. Each C—H bond is a σ bond from the overlap between an s orbital on H and an sp^3-hybridized orbital on C.

That leaves one more bond to take care of—the second bond of the C=C double bond. In Figure 3-22, you can see that that bond is the result of the overlap of the two unhybridized p orbitals, creating the π bond we previously saw in Figure 3-20. The C=C double bond is therefore comprised of a σ bond and a π bond. As you gain experience with double bonds between other atoms, such as a C=O, you will see that every double bond is comprised of both a σ and π bond.

Next, we examine HC≡CH, in which the C atoms have linear geometries. Both are therefore sp hybridized (Table 3-3). The two hybridized orbitals on each C point 180° apart, leaving two unhybridized orbitals that are perpendicular to each other (Figure 3-14). One hybridized orbital from each C points to the other C (Figure 3-23), resulting in a σ bond of the type in Figure 3-18. Each of the remaining hybridized orbitals overlaps with the s orbital of an H atom to create a σ bond, similar to that in Figure 3-19. Finally, the second and third bonds of the C≡C triple bond must be π bonds. Each π bond is the result of overlap between two unhybridized p orbitals (one from each C atom), as in Figure 3-20. The p_y orbital from one C atom overlaps with the p_y orbital from the other C to create one π bond. The second π bond is formed from the overlap of two p_z

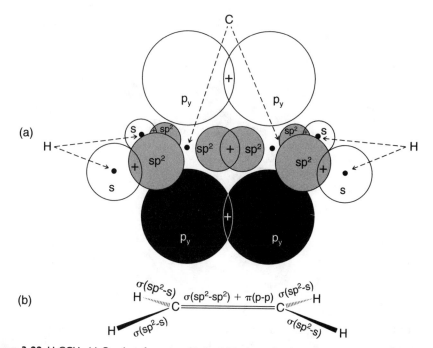

Figure 3-22 H_2CCH_2. (a) Overlap of atomic orbitals within the molecule to form the bonds. Hybridized orbitals from the C atoms are gray, and pure s and p atomic orbitals are black and white. "Good" overlap is indicated by "+". (b) Dash-wedge representation of the molecule. Each C—H bond is a σ bond resulting from the overlap between an s orbital on H and an sp^2-hybridized orbital on C. The C=C double is comprised of a σ bond and a π bond. The σ bond is from the overlap between two sp^2-hybridized orbitals. The π bond is from the overlap between two unhybridized p orbitals. The p orbitals are enlarged to explicitly show the overlap between them.

orbitals—one from each C atom. Therefore, a C≡C triple bond is comprised of one σ bond and two π bonds. Likewise, a triple bond between any two atoms is composed of one σ bond and two π bonds.

In our final example, we construct the MO picture of a formaldehyde molecule. Formaldehyde has two lone pairs of electrons on the O atom (Figure 3-24). The lone pairs are used to determine the hybridization on the atom, but, when constructing the MOs, they are dealt with last. First note that the C atom has three "groups" of electrons—two single bonds and a double bond. It is therefore sp^2 hybridized. The O atom also has three groups—a double bond and two lone pairs. It, too, is sp^2 hybridized. A σ bond is formed by the overlap of two sp^2 orbitals, one from C and one from O. That leaves two more sp^2 orbitals and an unhybridized p orbital on each atom. The p orbitals overlap to form a π bond, such that the C=O double bond is comprised of both a σ and a π bond. The two unused sp^2 orbitals on the C atom are taken care of by overlapping them with s orbitals from H atoms, forming two C—H single bonds.

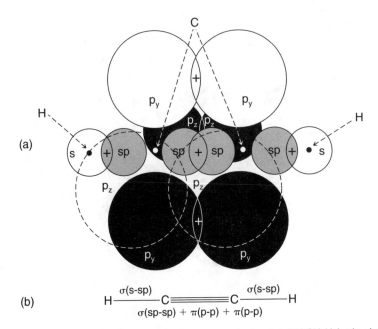

Figure 3-23 HCCH. (a) Overlap of atomic orbitals to create the bonds in HCCH. Hybridized orbitals are gray, and pure s and p atomic orbitals are black and white. "Good" overlap is indicated by "+". The unshaded lobes on each p_z orbital in front of the molecule are cut away (dotted circles) to show the details of the other bonds. The shaded lobes of the p_z orbitals are still shown behind the molecule. (b) The entire molecule is linear. Each C—H bond is a σ bond resulting from the overlap between an s orbital from H and an sp-hybridized orbital from C. The C≡C triple bond is comprised of one σ bond from the overlap of two sp orbitals, and two π bonds, each from the overlap of a pair of p orbitals.

There are still two sp^2 atomic orbitals on the O atom that have not been accounted for. Notice that neither of those orbitals overlaps with any other orbitals, and therefore remain unchanged in the molecule. Looking back at the Lewis structure, we also see two lone pairs of electrons on O that have not yet been accounted for. Each lone pair ends up occupying one of those sp^2 orbitals. That completes the MO picture of the $H_2C=O$.

3.9 APPLICATION: BOND ROTATION ABOUT σ AND π BONDS

In the three-dimensional representation of H_3C—CH_3 in Figure 3-21, one CH_3 group is drawn with a particular orientation relative to the second CH_3 group. Likewise, the representation of $H_2C=CH_2$ in Figure 3-22 is drawn with one CH_2 group in a particular orientation relative to the second. In H_3C—CH_3, it turns out that one CH_3 group can freely rotate relative to the other. In $H_2C=CH_2$, on the other hand, the molecule is "locked" with that orientation of the CH_2 groups.

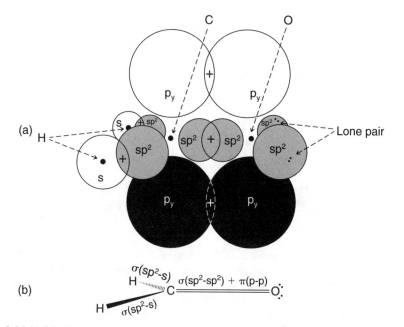

Figure 3-24 H_2CO. (a) Overlap of atomic orbitals to create the bonds in H_2CO. Hybridized orbitals are gray, and pure s and p atomic orbitals are black and white. "Good" overlap is indicated by "+". Lone pairs are shown in the two sp² orbitals on the right. (b) Dash-wedge notation of the molecule. Each C—H bond is a σ bond resulting from the overlap between an s orbital from H and an sp²-hybridized orbital from C. The C=C double bond is comprised of one σ bond from the overlap of two sp² orbitals, and a π bond from the overlap of a pair of p orbitals.

A CH_3 group can freely rotate because the bond that connects the two CH_3 groups—a σ bond only—is unaffected during rotation (Figure 3-25a). Recall that that σ bond is the result of the overlap between the lobes of two hybridized orbitals. Rotation of the CH_3 groups carries with it a rotation of the hybridized orbital that is part of the C—C single bond. However, due to the lobe's spherical nature, rotation of that hybridized orbital about the bonding axis does not change the amount of orbital overlap. For this reason, the bond itself remains unaffected. *In general, there is free rotation about any single bond*, given that a single bond is always comprised of one σ bond.

Similarly, when one CH_2 group in $H_2C=CH_2$ is rotated relative to the other, the σ bond of the C=C double bond is unaffected. This is not true for the π bond (Figure 3-25b). The rotation of a CH_2 group carries with it a rotation of the unhybridized p orbital on that C atom about the bonding axis. Rotating one p orbital relative to the other destroys the "good" overlap and effectively breaks the π bond. The energy cost associated with breaking a π bond is too great, which causes the CH_2 groups to be locked in place. *In general, then, groups cannot rotate about double bonds*, because all double bonds are comprised of one σ bond and one π, and the π bond locks the groups in place.

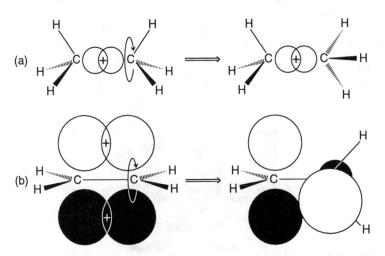

Figure 3-25 (a) Two different orientations of the CH$_3$ groups relative to one another during rotation about the C—C bond in H$_3$CCH$_3$. The overlap of the hybridized orbitals is unaffected by the rotation, meaning that such a rotation is allowed in the molecule. (b) Two different orientations of the CH$_2$ groups relative to one another during rotation about the C═C bond in H$_2$CCH$_2$. The overlap of the hybridized orbitals is unaffected. However, the "good" overlap between the p orbitals is destroyed, meaning that the π bond must be broken in order to accomplish bond rotation. Because breaking the π bond costs too much energy, rotation about a double bond is not allowed.

Notice that in order for the π bond in H$_2$C═CH$_2$ to remain intact, the plane defined by one CH$_2$ group must be parallel to the plane defined by the other. That requires all six atoms to be in the same plane. In general, *when two sp^2 C atoms are doubly bonded together, those two C atoms, along with the two other atoms on each sp^2 carbon (six atoms total) must all be in the same plane.* This is illustrated in Figure 3-26.

3.10 APPLICATION: CIS/TRANS PAIRS

Because bond rotation about a π bond is not allowed, it is possible for two different configurations to exist about a double bond, which are related by the *imaginary* rotation of 180° about the double. For such cases, we say that a **cis/trans pair** exists—one of the configurations is called a **cis** configuration, and the other is called a **trans** configuration.

Figure 3-26 A molecule of 2-methyl-2-heptene. The five C atoms and the H atom encompassed by the dotted rectangle must all be in the same plane.

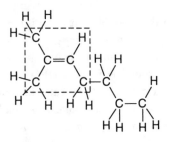

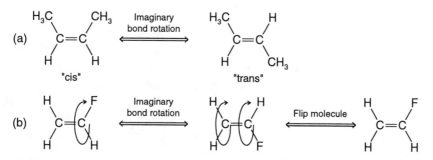

Figure 3-27 (a) Cis and trans pair of 2-butene. One is converted into the other by an *imaginary* rotation of 180° about the double bond. (b) Rotation of 180° about the double bond in H_2C=CHF results in exactly the same molecule. (left) The CHF group is rotated while the CH_2 is frozen, resulting in the structure in the middle. (middle) The entire molecule is flipped over, about the axis of the double bond, resulting in the structure on the right. (right) Same structure as that on the left. Therefore, a cis/trans pair does not exist for this molecule.

One of the simplest molecules that has a cis/trans pair is 2-butene, CH_3CH=$CHCH_3$ (Figure 3-27a). In one configuration, the two CH_3 groups reside on the same side of the double bond, and in the other configuration, the CH_3 groups are on opposite sides. Clearly, then, they are different molecules. Note in particular that an imaginary rotation of 180° about the double bond converts one configuration into the other.

In general, to determine whether a cis/trans pair exists for a double bond between two sp^2 C atoms, we must examine the four groups bonded to those C atoms—two groups on each end. *If each C that is part of the double bond has two groups that are different from each other, then a cis/trans pair exists for that double bond.* In CH_3CH=$CHCH_3$, for example, each C atom of the double bond is bonded to an H and a CH_3. These two groups, of course, are different, consistent with the fact that this molecule indeed has a cis/trans pair.

A molecule of H_2C=CHF, on the other hand, does not have a cis/trans pair, because the sp^2 C on the left possesses two identical H atoms. An imaginary rotation of 180° about the double bond therefore results in exactly the same molecule. Specifically, if the CH_2 group is rotated while the CHF group is frozen, the net result is simply the exchange of the two identical H atoms. Rotation of the CHF group, on the other hand, at first appears to result in a different molecule, but, in fact, does not (Figure 3-27b). If the entire species that results is flipped over, it is exactly the same species as that before the rotation about the double bond.

3.11 APPLICATION: LINE STRUCTURES FOR MOLECULES

In Chapter 2, we saw the utility of Lewis structures. However, Lewis structures can become quite tedious to draw for large molecules, particularly if every atom is included. A more convenient representation, which has no loss of structural information, is called the **line structure.** It invokes the octet rule, along with aspects of VSEPR theory.

Line structures are quick to draw because C atoms are not written explicitly, and H atoms bonded to C are not written at all. *C atoms are implied at the end of every bond, unless another atom is written there.* That includes the intersection of two bonds. The octet rule comes into play, because *enough H atoms are assumed to be bonded to each C atom to fulfill the octet (i.e., a total of four bonds).* VSEPR theory is invoked when drawing a string of C atoms bonded together. A string of tetrahedral C atoms, for example, is depicted as a zigzag structure, given that the C—C—C bond angle is bent (~109.5°). Similarly, the C—C—C bond angle of a trigonal planar C is depicted as bent (~120°). A linear C, of course, is depicted as linear.

Both the Lewis structure and the line structure of $CH_3CH_2CH_2CH_2CH_2CH_2CH_2NH_2$ are shown in Figure 3-28a. The intersection of each bond represents a C atom, as does the end of the bond on the molecule's left side. The NH_2 group is written in explicitly.

Figure 3-28 also shows the Lewis structures and the corresponding line structure for a variety of other molecules. Examples of a double and triple bond are shown in (b) and (c), respectively. Examples of non-C/H atoms are shown in (d) and (e). In Figure 3-26f and 3-26g, the formal

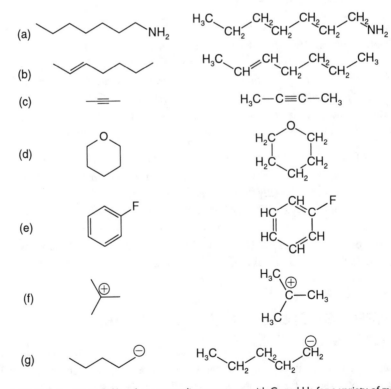

Figure 3-28 Line structures (left) and corresponding structures with Cs and Hs for a variety of molecular species. (a) 1-aminoheptane. (b) trans-2-heptene. (c) 2-butyne. (d) tetrahydropyran. (e) fluorobenzene. (f) t-butyl cation. (g) 1-pentyl anion.

charge implies that the C atom forms only three bonds. In the case of the cation, there are three bonds and no lone pairs. In the case of the anion, there are three bonds and one lone pair. If the lone pair is not shown explicitly in the line structure of the anion, the -1 formal charge implies that it is there.

Problems

3.1 Identify the hybridization on every nonhydrogen atom in each of the following species. Note: Lone pairs are not shown and can be assumed.

(a) $H_3C—OH$

(b)

(c)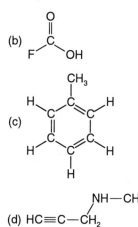

(d) $HC\equiv C—CH_2$ with NH—CH₃ substituent

3.2 Which molecule has a σ bond between an sp orbital and an sp² orbital?

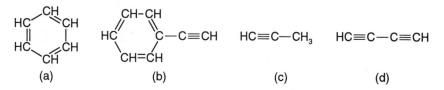

(a)   (b)   (c)   (d)

3.3 Circle all of the following species that have a cis/trans pair.

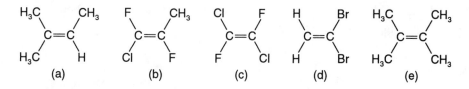

(a)   (b)   (c)   (d)   (e)

3.4 Draw the molecular orbital picture (similar to Figures 3-21 to 3-24) for $:C\equiv O:$.
3.5 Draw the molecular orbital picture (similar to Figures 3-21 to 3-24) for $H_2C=NH$.

67

3.6 (a) Draw the molecular orbital picture (similar to Figures 3-21 to 3-24) for allene ($H_2C=C=CH_2$). (Hint: First draw the central C atom, with all of its orbitals. Then draw the C atom of the CH_2 group on the left and form the C=C double bond. Do the same for the C atom on the CH_2 group on the right, realizing that certain orbitals from the central C atom have already been used up. Finally, construct the C—H single bonds.) (b) The MO bonding picture of allene dictates whether or not the two Hs on the left are in the same plane as those on the right. Are they in the same plane or in a different plane? What is it that specifically dictates the orientation of one pair of Hs relative to the other pair?

3.7 Must all of the atoms in the following molecule be in the same plane? Why or why not?

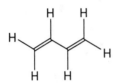

3.8 Draw the line structure for each of the following species.

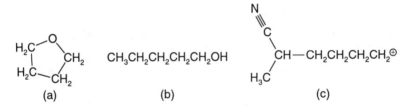

(a) (b) (c)

3.9 For each line structure, write the corresponding structure with all C atoms, H atoms, and lone pairs of electrons included explicitly.

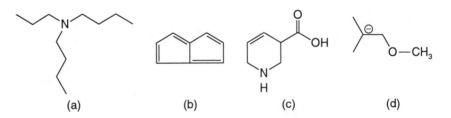

(a) (b) (c) (d)

3.10 In Problem 3.9d, what is the electronic and molecular geometry about the C atom bearing the negative charge?

Isomerism

4.1 INTRODUCTION

Two molecules are **isomers** of each other if they are *different molecules that have the same molecular formula*. This is a rather straightforward definition and a straightforward concept. In fact, you have probably worked with, and were comfortable with, isomers and isomerism in general chemistry. However, isomerism is one of the first concepts that many students struggle with in organic chemistry. Why?

Part of the answer is that there are several different types of isomers with which we must be concerned. That is, molecules that have the same formula can actually differ in any of several ways (discussed in detail later). If you encountered isomers in general chemistry, you probably dealt only with one main branch of isomers called **constitutional isomers,** which differ from each other in a specific way. However, throughout this chapter we will also discuss another main branch of isomers, called stereoisomers, which are further categorized into enantiomers and diastereomers.

Also, many students struggle with isomerism because of its connection to chemical reactivity. Two molecules that are isomers of each other may appear to be almost identical but could behave quite differently. One isomer can be used as antinausea medication for pregnant women, while the other may cause birth defects in the fetuses of those same pregnant women (if interested, search on "thalidomide"). On the other hand, two molecules that appear to be structurally very different could have nearly identical chemical reactivity.

This chapter's first goal is to provide you with a solid understanding of the various ways in which two molecules can be isomers of each other. Simply keeping the definitions straight can often be quite difficult. The second goal is to teach you how to determine the precise relationship between any two molecules, using a very systematic method. Are they actually isomers of each other, or are they the same molecule drawn differently? If they are isomers, what type? Finally, we will examine the importance of isomerism in chemical reactions—something you will undoubtedly be held accountable for on your organic chemistry exams.

4.2 ISOMERS: A RELATIONSHIP

The confusion with isomers and isomerism can begin with a lack of understanding of what an isomer means. Many students can recite the definitions of the various isomer types but will try to look at a molecule and say, "That's an enantiomer" or "That's a diastereomer." Although we haven't covered the specific definition of *enantiomer* or *diastereomer,* you can be assured that such statements make no sense whatsoever. This is because *any type of isomerism is a relationship between two molecules.* It does not describe a single molecule. Therefore, each of those statements is analogous to pointing at someone and saying, "There's a cousin."

A statement that makes more sense using the word *cousin* would be, "Those two girls are cousins of each other," or, "She is that man's cousin." Likewise, using the word *enantiomer, diastereomer,* or even *isomer,* makes no sense unless you are talking about two or more molecules. It is okay to say, "Those two molecules are enantiomers of each other," or "This molecule is a diastereomer of that one." Keep this in mind as you go through these sections (in particular, the section on stereochemistry) and things will be less confusing.

4.3 CONSTITUTIONAL ISOMERISM

As mentioned briefly, constitutional isomers are the first of two main branches of isomers—the other being stereoisomers (Figure 4-1). In order to be constitutional isomers of each other, two molecules must first be isomers of each other; that is, they must have the same molecular formula, and they must be different molecules. *Constitutional isomers differ in their connectivity.*

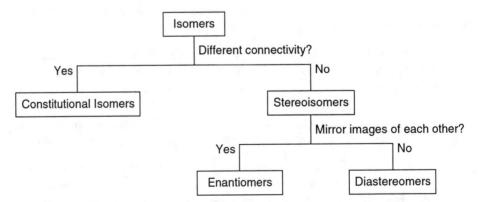

Figure 4-1 Flowchart showing the relationships between various types of isomers.

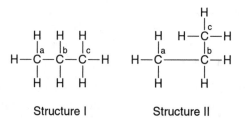

Structure I   Structure II

Figure 4-2 Both Structure I and Structure II clearly have the same molecular formula of C_3H_6, but their Lewis structures are drawn differently. Nevertheless, they have the same connectivity and are therefore not constitutional isomers of each other.

Before we move on, we must ensure that we have a solid understanding of connectivity. Put simply, **connectivity** is *the way in which a molecule is connected together,* which includes (1) the types atoms, and (2) the types of bonds. Nothing less, nothing more. Connectivity specifically does not include anything about the three-dimensional shape. Therefore, the Lewis structure conveys all information about a molecule's connectivity. To better illustrate this point, we provide some examples.

Let's begin with a trivial example. In Figure 4-2, there are two Lewis structures of molecules with the formula C_3H_8. They are drawn differently, but their connectivity is identical. In both Structure I and Structure II, there is a C atom labeled "a" that is singly bonded to three H atoms and to one C atom labeled "b." The C atom labeled "b" is singly bonded to two H atoms and to a C atom labeled "c." The C atom labeled "c" is singly bonded to three more H atoms. Therefore, even though the Lewis structures are drawn differently, they have identical connectivity. Consequently, they cannot be called constitutional isomers of one another.

Let's now look at another example. Figure 4-3 contains two Lewis structures, both with the same molecular formula of C_8H_{18}. In this case, the two structures are indeed constitutional isomers, because they differ in connectivity. To prove this, you need find only one point

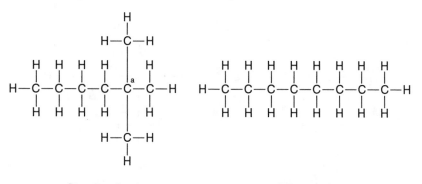

Structure I   Structure II

Figure 4-3 Structure I and Structure II bear the same molecular formula and differ in their connectivity. They are therefore constitutional isomers of each other.

of difference in the connectivity between the two molecules. Notice that in Structure I in Figure 4-3, there is a C atom, labeled "a," that is attached by single bonds to four other C atoms. In Structure II, there is no C atom like this. There are other points of difference, but we have already found our one.

In yet another example, let's look at Figure 4-4, which again contains two Lewis structures having the molecular formula C_8H_{18}. Are these structures constitutional isomers of each other? No. But at first glance it may appear as if they are. If you thought that they were constitutional isomers, this is probably because you did not attempt the problem in a systematic fashion. Most likely, you just looked at the overall picture of each molecule, were convinced that they *looked* different, and concluded that they must be constitutional isomers. This will get you into trouble!

To approach a problem like this in a systematic fashion (which I encourage you to do EVERY time), use the "two finger method." Pick any atom in Structure I, and point to it with your left index finger. Let's choose the C atom labeled "a." It is bonded to three H atoms and to a C atom that is bonded to two more C atoms—one is part of a CH_3 group, and the other is part of a CH_2 group. The C atom labeled "z" in Structure II shares these same features. Keeping your left index finger where it is, use your right index finger to point to the C atom labeled "z." Now, move your left index finger over to the next C atom, labeled "b." This is bonded to an H atom, a CH_3 group on the left, a CH_3 group above, and a CH_2 group on the right. Next move your right index finger over one C atom to the left, labeled "y." Does the "y" carbon in Structure II share those same features as the "b" carbon in Structure I? Yes. So, thus far, we have not run into any points of difference. One more time, so you get used to it, move your left index finger over one more C atom to the right, labeled "c," and move your right index finger over one C atom to the left, labeled "x." Do the "x" and "c" C atoms share the same features in connectivity? Again, yes. Both of these

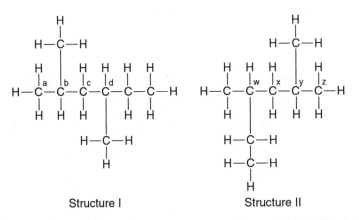

Structure I   Structure II

Figure 4-4 Structure I and Structure II both have the same molecular formula of C_8H_{18}. They furthermore have exactly the same connectivity and are therefore not constitutional isomers of each other.

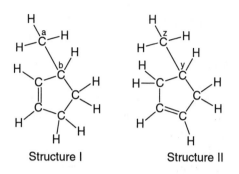

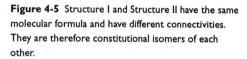

Figure 4-5 Structure I and Structure II have the same molecular formula and have different connectivities. They are therefore constitutional isomers of each other.

Structure I   Structure II

C atoms are bonded to two H atoms and to two C atoms that are each bonded to a CH_3 group. We could continue like this all the way down the chain of C atoms (which I encourage you to do, starting with comparing carbon "d" and carbon "w"), and we would find that at each point, the C atoms have the identical connectivity. Therefore, there is no point of difference in connectivity between these two molecules.

Let's look at another example, which brings multiple bonds to the table. There are two Lewis structures in Figure 4-5, each of which has a double bond, and each of which has the formula C_6H_{10}. As an added twist, each also has a five-carbon ring. As before, the safest way to test whether these are constitutional isomers is to use both index fingers. With your left index finger, pick any atom in Structure I, say, the C labeled "a," which is the C of the CH_3 group. It has three H atoms and one C atom (labeled "b") singly bonded to it. There is an atom in Structure II that seems to have the same connectivity—the C atom labeled "z." Put your right index finger on it. Now move your left index finger over to atom "b," which is a C atom that is singly bonded to three carbon atoms and to an H atom. Move your right index finger over to the C atom labeled "y," which is also a C atom that is singly bonded to three other C atoms and to an H atom. However, atom "b" and atom "y" are different C atoms—one of the C atoms that "b" is bonded to is doubly bonded to another C atom, whereas all of the C atoms that "y" is bonded to contain only single bonds. This is a subtle difference between the connectivities of atoms "b" and "y," but one that clearly makes them different atoms. Therefore, since we have found our point of difference in connectivity, it must be that Structures I and II are constitutional isomers of each other.

4.4 STEREOISOMERISM

Stereoisomerism is the second major branch of isomerism (Figure 4-1). Since there are only two branches, a pair of **stereoisomers** can be recognized as *two isomers that are NOT constitutional isomers*. In other words, stereoisomers must have the same molecular formula, and they must also have identical connectivity. To be isomers of each other, however, they must be different in some way. And if both molecules have the same connectivity (same atoms and same type of bonding), then the only way that the two molecules can be different is if the atoms are *arranged differently in space.* Figure 4-1 suggests that there are two ways in which this can occur. One way results in *enantiomers,* and the other results in *diastereomers.*

Before we look at some examples of enantiomers and diastereomers, let's first define these two relationships. Both are defined by comparing a molecule with its mirror image (we will practice drawing mirror images of molecules later). **Enantiomers** are *mirror images of each other that are not superposeable.* In other words, two molecules are enantiomers of each other if (1) they are mirror images of each other, and (2) they are *different molecules* (if they are superposable, then every atom and bond can be overlaid perfectly, and they are exactly the same molecule—they cannot be isomers). On the other hand, **diastereomers** are *stereoisomers that are not mirror images of each other.*

Let's now look at some examples to better see what gives rise to stereoisomerism and to understand how to determine whether a given pair of stereoisomers are enantiomers or diastereomers. The first example of a pair of stereoisomers is the most straightforward and involves a relationship that we discussed in Chapter 3—cis/trans isomerism. Let's look at cis and trans-1,2-dicholorethene, Cl—CH=CH—Cl (Figure 4-6). The **trans isomer** is that in which *the two non-H atoms are on opposite sides of the* C=C, and the **cis isomer** is that in which *the two non-H atoms are on the same side.*

To convince ourselves that a cis/trans pair constitutes stereoisomers, we must first establish that the two are actually isomers—that is, that they have the same molecular formula, but are *different molecules.* We quickly see that they do have the same molecular formula of $C_2H_2Cl_2$. And the double bond ensures that they are different molecules, because, as we learned in Chapter 3, there is no free rotation about a double bond (converting between the cis and trans forms would otherwise involve a rotation of 180° about the double bond, which requires breaking the π bond).

We must now prove that these isomers are stereoisomers. Going back to the definition of stereoisomers, this entails showing that the two molecules have the same connectivity, which can be done in one of two ways. We have done the first way before, looking at the specific connectivity of each atom in the molecule (using both index fingers to keep track). The second is to take advantage of the fact that a hypothetical rotation of 180° about the double bond would convert between the cis and the trans forms but would not change the connectivity within the molecule. Therefore, yes, these molecules must have the same connectivity and must be stereoisomers of each other.

To determine which type of stereoisomers these molecules are—enantiomers or diastereomers—we need only ask ourselves whether they are mirror images of each other or not.

Figure 4-6 Structure I is the trans isomer of 1,2-dichloroethene, and Structure II is the cis isomer.

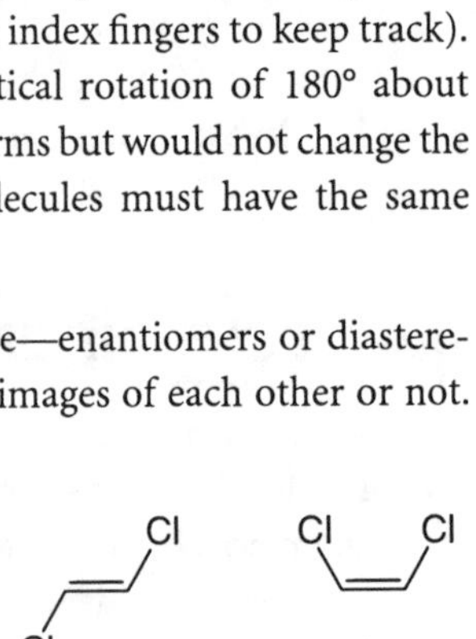

Structure I   Structure II

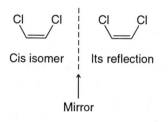

Figure 4-7 The cis isomer of 1,2-dichloroethene and its mirror image. Here, the dashed line represents a mirror through which we are taking the reflection.

Cis isomer | Its reflection

↑
Mirror

Although the answer to this question may be obvious, the cis isomer and its mirror image are shown in Figure 4-7. The mirror image of the cis isomer is also the cis isomer—that is, the mirror image is itself. Therefore, the cis and trans isomers are not mirror images of each other, which makes them diastereomers of each other.

Given the importance of mirror images in stereoisomerism, we should practice drawing mirror images of molecules in general. In the previous example, drawing the mirror image of cis-1,2-dichloroethene is easy, but that is because the molecule is only two-dimensional and lies in a single plane. Most molecules are three-dimensional, so it is best to *use a systematic method in which to draw mirror images.*

The first step is to draw the molecule of interest, with the appropriate level of detail. As a nice example of a three-dimensional molecule, let's choose CH_2FCl (Figure 4-8). Next, draw a

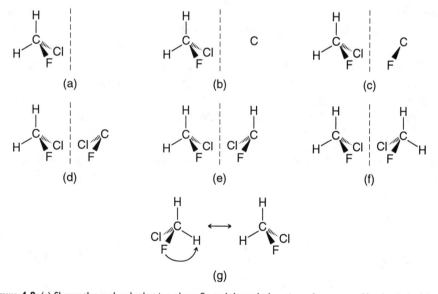

Figure 4-8 (a) Shows the molecule that is to be reflected through the mirror (represented by the dashed line). (b) The result of having reflected the C atom through the mirror. (c) The result of having reflected the F through the mirror. (d) The result of having reflected the Cl through the mirror. (e) The result of having reflected one H atom through the mirror. (f) The result of having reflected the second H atom through the mirror, completing the reflection. (g) A rotation of the molecule's reflection in space shows that it is the same as the original molecule.

dashed line adjacent to the molecule—this represents the mirror through which the molecule you drew will be reflected. In Figure 4-8a, the molecule is drawn in three dimensions, and the "mirror" is drawn vertically on its right. Now, we must reflect each atom, one by one, through the mirror. Let's start with the C atom. In the original molecule, the C atom is drawn in the plane of the paper, roughly toward the mirror's middle. In the reflection, it should also be in the plane of the paper, the same height up the mirror, and the same distance from the mirror as it is in the original molecule. The result is Figure 4-8b. Next we can reflect the F atom through the mirror. In the original molecule, the F atom is in front of the plane of the paper (that is, pointing toward you), is toward the bottom of the mirror, and is closer to the mirror than the C atom was. All of this must also be true in the mirror image. The result is Figure 4-8c. Notice that the bond connecting the C and F atoms in the original molecule is a wedge, indicating that the F atom is in front. Likewise, the C—F bond in the mirror image is also a wedge. Figure 4-8d is the result of having reflected the Cl atom through the mirror. And reflecting the two H atoms completes the reflection (Figures 4-8e and 4-8f).

What is the relationship between the original molecule and the mirror image in Figure 4-8? They have the same molecular formula and the same connectivity, since taking a reflection through a mirror does not change either of these (if you don't believe this, then do the test with both index fingers!). Given that they are mirror images of each other, are they then enantiomers of each other? No. This is because the mirror image is superposable on the original molecule—that is, the original molecule is identical to its mirror image. Whether or not you have difficulty rotating three-dimensional pictures in your head, the best thing to do to convince yourself of this is to *build the original molecule and its mirror image as two separate molecules using a molecular modeling kit* (I highly suggest that you do this). You can rotate the mirror image's molecule in space and line it up perfectly with the original molecule. Every atom will line up—the Fs, the Cls, the Cs, and all of the Hs. If, however, you can easily rotate three-dimensional objects in your head, you should realize that if you rotate the mirror image by roughly 120° about the axis that points in and out of the paper (Figure 4-8g), the result is the original molecule. Therefore, the original molecule is the same as its mirror image.

Finally, let's look at CHFClBr. It is shown with its mirror image in Figure 4-9 (convince yourself using the systematic method of reflecting one atom at a time). In this case, what is

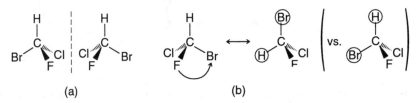

(a)   (b)

Figure 4-9 (a) Three-dimensional representation of CHFClBr and its mirror image. (b) The mirror image is rotated in space, so that three of the five atoms (C, F, and Cl) line up with atoms in the original molecule (in parentheses), but H and Br (circled) do not.

the relationship between the original molecule and its mirror image? Once again, build each of the molecules using a molecular modeling kit and physically rotate them in space to see if they match up completely. You will find that they do not. The best that you can do is line up three of the five atoms, and the other two will not line up. This is depicted in Figure 4-9, where C, F, and Cl line up perfectly in the original molecule and in the mirror image, but H and Br do not.

A term that goes hand in hand with stereoisomerism is *chirality*. A molecule is said to be **chiral** (pronounced *kiral*) if it *has an enantiomer*. This definition is straightforward enough, but remember that, as with all of the fundamental concepts covered in this book, you must be able to go beyond simple recitation of the definition. You must be able to apply it and therefore understand it well.

Before we go through some examples, let's look at the difference between the definition of *chiral* and the definition of *enantiomer*. Keeping these definitions straight can be frustrating, given that one is used in the definition of another, but it is easy if you remember that *enantiomer* is a *relationship between two molecules. Chirality* is not—it is a word that *describes a single molecule*. A single molecule can be chiral or **achiral** (that is, not chiral), but a molecule can only be an enantiomer *of another molecule*.

So how do we determine whether a molecule is chiral or achiral? Start by taking a closer look at the original molecule in Figure 4-9, because it is chiral—its mirror image is different from itself (in fact, both the original molecule and its mirror image are each chiral). This molecule is not special. As it turns out, *a molecule must be chiral if it has exactly one tetrahedral atom bonded to four different atoms or groups*. You can prove this by replacing the four atoms in Figure 4-9 with any four groups that are all different from each other— say, a CH_3 group, a CH_2CH_3 group, an OH group, and an NH_3 group. If you then use a molecular modeling kit to build this molecule and its mirror image, you will find that they are not superposable. As in Figure 4-9, if you attempt to line up the molecule with its mirror image, the best you can do is line up the central C atoms and two of the remaining four groups. The other two groups will not line up.

It should be clear that a tetrahedral atom bonded to four different groups of atoms is important in stereoisomerism. Consequently, such atoms are given their own names—they are called **stereocenters** or **chiral centers.** We can therefore reword the previous statement to say that *a molecule that contains exactly one stereocenter must be chiral.* (As we will see shortly, a molecule with two or more stereocenters may be chiral but does not have to be.)

Next we can examine the two molecules that were achiral (Figures 4-7 and 4-8) in order to gain some insight into why they are achiral. What these molecules share in common is that they each have a **plane of symmetry,** or a **mirror plane.** That is, there is a way to divide each molecule in two, such that one half is the mirror image of the other half. For example, the molecule in Figure 4-7 lies entirely in a single plane. That plane is the molecule's plane of symmetry. The top half of each atom can be considered the mirror image of the bottom

Figure 4-10 The plane of symmetry of CH_2ClF. The plane contains C, F, and Cl and is perpendicular to the plane of the paper. The dotted line represents the line bisecting the H—C—H angle.

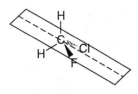

half—or, the mirror image of these atoms are themselves. The plane of symmetry in the molecule in Figure 4-8, on the other hand, is that which contains the C, F, and Cl atoms and bisects the H—C—H angle (Figure 4-10). The mirror image of the C, F, and Cl atoms are themselves, and the mirror image of one H atom is the other H atom. The lesson: *Any molecule that has a plane of symmetry must be achiral.*

Let's now step it up a bit and examine molecules that have more than one stereocenter. Figure 4-11a, for example, shows a molecule that contains two stereocenters—each C of the ring that is bonded to a CH_3. To be sure that those two C atoms are stereocenters, we must ensure that each is bonded to four different groups. Clearly the H atom and CH_3 group are different from each other, and the two groups that are part of the ring are different from the H and CH_3. But are the groups that comprise the ring different from each other? To answer this, we need only apply the connectivity test we learned while working with constitutional isomers. Begin by placing your left and right index fingers on one of the C atoms we think might be a stereocenter. Move your left index finger one carbon atom to the left around the ring, and move your right index finger one carbon atom to the right. You will notice that those two C atoms have different connectivities. One is bonded to two C atoms that are part of the ring and to two H atoms. The other is bonded to two C atoms that are part of the ring, to one H atom, and to one CH_3 group. Consequently, each C atom that is bonded to a CH_3 group is bonded to four different substituents (indicated by the number 1-4) and is indeed a stereocenter.

Another way to convince yourself that the two groups on the ring (labeled 3 and 4 in Figure 4-11a) are different is to imagine standing at the C atom in question (bonded to a H atom and to CH_3 group). As you face the center of the ring, you'll notice that the ring looks different on your left than it does on your right. Looking to your right, you would see a C atom of the ring that is bonded to a CH_3 group. To your left, there is no C atom of the ring bonded to a CH_3 group.

Figure 4-11 (a) Explicity shows that the four groups attached to the C on the ring are different. Therefore, that C on the ring is a stereocenter. (b) The vertical dashed line is a plane of symmetry. The left half of the molecule is the mirror image of the right half. Therefore, this molecule is achiral.

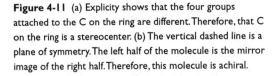

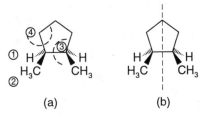

(a)   (b)

Is that molecule chiral or achiral? There are two ways to determine this. The foolproof way (which you can *never* practice enough) is to draw the reflection alongside the original molecule, build both using a molecular modeling kit, and then try to line them up. The second way is to apply the plane of symmetry test. It turns out that the molecule in Figure 4-11 does have a plane of symmetry, which is depicted as a dashed line in Figure 4-11b. Therefore, that molecule must be achiral. It is also said to be **meso,** meaning that it is *a molecule that contains at least one stereocenter, but overall is achiral.*

A similar molecule is shown in Figure 4-12; it, too, possesses two stereocenters. If you build that molecule and its mirror image using a molecular modeling kit, you will find that there is no way to perfectly line the two up. Likewise, take the time to search for a plane of symmetry in the molecule. You will not find one. Therefore, this molecule is chiral.

What about the relationship between the two molecules in Figures 4-11 and 4-12? They each have the same molecular formula and the same connectivity. Furthermore, they are *different* molecules. In one molecule, both CH_3 groups are on the same side of the ring (i.e., both are pointing toward you), and in the other molecule, they are on opposite sides (if you cannot immediately see that those molecules are different, build each molecule with the kit and try to line them up). Consequently, these molecules are stereoisomers of each other, meaning that they must be either enantiomers or diastereomers of each other. To test which one, we must determine whether they are mirror images of each other or not. It turns out that they are not mirror images of each other (again, if you cannot immediately see, go through the complete exercise of drawing each molecule's mirror image), which makes them diastereomers.

You have just seen a second example of a pair of diastereomers. The first was a pair of planar molecules containing no stereocenters (cis/trans isomers). This one is three dimensional, containing more than one stereocenter. The lesson here is that *molecules that have more than one stereocenter can have potential enantiomers, and they can have potential diastereomers.* It is also possible for such a molecule (i.e., with two or more stereocenters) to have a plane of symmetry, making it achiral and therefore meso.

Regarding things that trip students up, here are a few final notes about stereochemistry. The first is with the definition of a stereocenter. Even though there is no mention of a C atom in the definition, students tend to assume that in order to be a stereocenter, the tetrahedral atom must be C. However, this is not true. The N atom is another stereocenter

Figure 4-12 This molecule contains two chiral centers, marked by an *. It does not have a plane of symmetry, and is therefore chiral. If you build this molecule and its mirror image using a molecular modeling kit, you will find that they do not line up and are different molecules.

that one can encounter in organic chemistry. It can be bonded to four different groups, making it a tetrahedral atom. It just turns out that an N atom with four different groups also bears a + charge. Second, students tend to assume that since the existence of stereo-centers within a molecule has a tendency to make a molecule chiral, then a molecule that has no stereocenters must be achiral. This is not true. You have a practice problem that gets at the heart of this idea.

4.5 PHYSICAL AND CHEMICAL BEHAVIOR OF ISOMERS

In the opening section of this chapter, we mentioned that in some cases, isomers could behave quite similarly, and in other cases, they can behave quite differently. Much of this has to do with the specific type of isomerism that two molecules share. Are they constitutional isomers or stereoisomers? If they are stereoisomers, are they enantiomers of each other or are they diastereomers?

Two molecules that are *constitutional isomers of one another must have different chemical properties* (what products are formed, reaction rate, equilibrium constants, etc.) *and physical properties* (boiling point, melting point, water solubility, etc.; see Chapter 7). The reason is that, by definition, they must have different connectivities, which, in turn, means that they must have different bonding in some way. For example, it may be that one of two constitutional isomers has a C=O, whereas the second molecule has a C=C. It may be that one of the molecules contains a ring in the structure, while the other does not.

How differently constitutional isomers behave largely depends on how different their connectivities are. Figure 4-13a shows two constitutional isomers of C_5H_{10}. Both structures are composed of one C=C double bond, and the rest are C—H and C—C single bonds. The structural difference between them is not great—the double bonds are simply found at different locations within the molecules. Not surprisingly, these two molecules behave quite similarly, both in their physical and chemical characteristics. On the other hand, Figure 4-13b shows two constitutional isomers that have quite different physical and chemical behavior. This should not be a surprise, given their differences in connectivity–e.g., one has a C≡N triple bond, whereas the other has nothing of the sort.

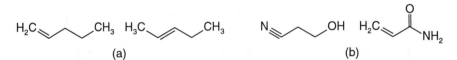

(a)   (b)

Figure 4-13 (a) Constitutional isomers of C_5H_{10}. The only difference is in the placement of the double bond along the carbon chain, suggesting that the two molecules should have quite similar chemical behavior. (b) Constitutional isomers of C_3H_5NO. They are constructed from different types of bonds, suggesting that they might have very different chemical behavior.

The notion of different connectivity leading to different physical and chemical behavior is expanded upon in Section 4.7.

Two molecules that are enantiomers of each other must behave EXACTLY the same in the presence of achiral species—that is, in an **achiral environment.** The reason is that enantiomers, even though their atoms are arranged differently in space, are mirror images of one another. To better understand this, let's use an analogy of feet and socks. Your left and right feet are chiral—one is the mirror image of the other, and they are not superposable. A sock, however, is achiral, because it has a plane of symmetry. From experience we know that the same sock will fit equally well on either foot. That is, each of those enantiomers "interacts" in exactly the same way with the achiral sock. Both enantiomers behave the same in the achiral environment of a sock.

On the other hand, *enantiomers behave differently in the presence of chiral species*—that is, a **chiral environment.** We can better understand this by considering the "interaction" between feet and shoes. We already know that your left and right feet are chiral. For the same reason, left and right shoes are chiral—they are enantiomers of each other. Again, we know that a left shoe does not fit on a right foot, and vice versa. That is, each of those enantiomers behaves differently in the presence of chiral shoes. Or, each enantiomer (your left or right foot) behaves differently in the chiral environment we are calling a shoe.

Finally, a *pair of diastereomers must behave differently.* This is readily seen with the cis and trans isomers of 1,2-dichloroethene, $Cl—CH=CH—Cl$ (Figure 4-6). Notice that there is no net dipole in the isomer with both Cl atoms on opposite sides of the double bond, whereas there is a considerable net dipole with the Cl atoms on the same side of the double bond. As we will see in Chapter 7, this translates into significant differences in physical properties and can be extended to differences in chemical reactivity.

Sometimes diastereomers can have significant differences in behavior, as with the cis and trans isomers of $Cl—CH=CH—Cl$. Other times, it may be that those differences are quite small. For example, the pair of diastereomers in Figures 4-11 and 4-12 are structurally very similar and would be expected to have nearly the same properties. However, no matter how similar diastereomers appear, their behavior differs in some way.

4.6 APPLICATION: INDEX OF HYDROGEN DEFICIENCY (DEGREE OF UNSATURATION)

A molecule's **index of hydrogen deficiency (IHD),** also known as its **degree of unsaturation,** is quite useful as an aid in determining the structure of an unknown compound. Knowing nothing more than the molecular formula, you can determine how many double bonds, triple bonds, or rings can possibly exist. Having such information can drastically reduce the number of candidates for the molecular structure. Combining the IHD with other information, such as that obtained from spectroscopy experiments (often given to you in the problem), usually can lead to clear identification of the compound.

The IHD for a given molecule can be defined as *the number of H_2 molecules absent from an analogous completely "saturated" molecule.* And a molecule is said to be **saturated** (with H atoms) if it has *the maximum possible number of H atoms.* For example, a molecule of ethane, H_3C—CH_3, is a saturated molecule, because it is not possible for a molecule with two C atoms to possess any more than six H atoms. Each C atom is singly bonded to four other atoms (which is the maximum possible, given the octet rule), three of which are H atoms. On the other hand, a molecule of ethene, H_2C=CH_2, has two fewer H atoms, or one fewer H_2 molecule, than the completely saturated two-carbon molecule. Therefore, its IHD = 1. Another H_2 molecule can be removed, yielding HC≡CH; such a molecule therefore has an IHD = 2. As a general rule, then, it appears that *each double bond contributes to an IHD of 1, and each triple bond contributes to an IHD of 2.*

To demonstrate this, we can examine CH_3CH=$CHCH$=CHC≡CH. There are two double bonds, each contributing 1 to the IHD, and there is one triple bond, contributing 2 to the IHD, for a total IHD = 4. Another way to reach the same number is to compare this seven-carbon molecule to a completely saturated seven-carbon molecule, such as $CH_3(CH_2)_5CH_3$, which contains sixteen H atoms. The difference is eight H atoms, or four H_2 molecules, consistent with an IHD = 4.

In addition to double and triple bonds, *each ring adds 1 to the IHD of a molecule.* The cyclic seven-carbon molecule, cycloheptane (Figure 4-14), has a molecular formula of C_7H_{14}. Compared to the completely saturated seven-carbon molecule previously mentioned, which has a formula of C_7H_{16}, there is a single H_2 molecule that is missing, for an IHD = 1. In a molecule that contains double bonds, triple bonds, and rings, determining the total IHD simply involves adding the IHDs contributed by each. For example, benzene, which is a cyclic six-carbon molecule that contains three double bonds (Figure 4-14), has a total IHD = 4. The same result can be obtained by comparing the formula of benzene, C_6H_6, to the analogous completely saturated (open-chain) molecule, which would have a formula of C_6H_{14}. Benzene is therefore missing eight H atoms, for an IHD = 4.

The IHD is most useful when given nothing more than the molecular formula of a molecule. From the molecular formula, you can determine the IHD, and with the IHD, you can subsequently determine the number of double bonds, triple bonds, or rings that could be

Figure 4-14 (a) Structure of cycloheptane, C_7H_{14}.
(b) Structure of benzene, C_6H_6.

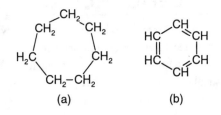

(a)   (b)

present, or the combination thereof. This works because *constitutional isomers have the same IHD as one another*. So, if you were given a molecular formula of C_7H_6, then by comparing it to the completely saturated molecule of C_7H_{16}, you can quickly deduce that the IHD = 5. A molecule that contains five double bonds is therefore consistent with the formula C_7H_6. So is a molecule that contains three double bonds and one triple bond, or a molecule that contains two rings, a double bond, and a triple bond. All of these examples would be constitutional isomers of one another.

The situation becomes slightly more tricky when the molecular formula you might be given contains atoms other than just C and H. It may contain O, N, F, Cl, etc. If you are asked to determine the IHD for such a molecule, draw a completely saturated molecule that contains the same number of each of the non-H atoms given in the molecular formula. Then simply add in as many H atoms the octet rule will allow and determine how many H_2 molecules are missing from the formula you are given.

Let's determine the IHD of $C_7H_7NO_2$. Step one is to draw a Lewis structure of a completely saturated molecule with seven Cs, one N, and two Os. The completely saturated molecule should contain all single bonds and no rings. If we connect all of the nonhydrogen atoms together with single bonds, and then add enough H atoms to complete the octets, we might come up with the structure shown in Figure 4-15, which has the formula $C_7H_{17}NO_2$. Whether or not such a molecule exists, it is completely saturated. Comparing this to the molecular formula we were given, it appears that the molecule we were given is missing 10 H atoms, for an IHD = 5.

This method also works with ions. Let's determine the IHD of $C_3H_3O_2^-$. In coming up with a completely saturated species, we can choose to put the negative charge on either a C or an O—the formula will be the same. Let's put it on an O atom, which means that that O atom will have three lone pairs of electrons and one covalent bond. If we connect the three C atoms in a row, followed by the two oxygen atoms, we end up with the following as a completely saturated species: $CH_3CH_2CH_2OO^-$ (the oxygen atoms have only single bonds). Its molecular formula is $C_3H_7O_2^-$, so that the molecular formula we are examining, $C_3H_3O_2^-$, is short four H atoms, for an IHD = 2.

4.6a Possible Pitfalls

Many organic chemistry textbooks present formulas or equations that can be used to determine the number of H atoms in a completely saturated molecule containing n carbon

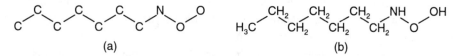

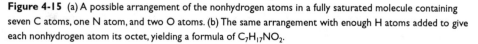

(a) (b)

Figure 4-15 (a) A possible arrangement of the nonhydrogen atoms in a fully saturated molecule containing seven C atoms, one N atom, and two O atoms. (b) The same arrangement with enough H atoms added to give each nonhydrogen atom its octet, yielding a formula of $C_7H_{17}NO_2$.

atoms. Whereas these equations can be quite convenient, each formula represents another thing to memorize. Trouble arises because the formula is different depending upon whether there are only Cs and Hs, or whether it contains Fs, Ns, Os, etc. For example, a completely saturated molecule that contains n Cs will possess 2n+2 H atoms. For each F atom the molecule contains, there will be one less H atom in the analogous completely saturated molecule. For each N atom, there will be one additional H atom. And each O atom does not change the number of H atoms in the analogous saturated compound. Furthermore, working with charged species requires additional adjustments to those formulas!

Needless to say, memorizing these formulas will serve to confuse you more than help you. Instead, I believe the best method for determining the formula for the completely saturated molecule is that which was done earlier for $C_7H_7NO_2$—a method that is simple to apply and involves no memorization. This method is independent of the types of bonds contained in the molecule and is independent of the charge the species carries.

4.7 APPLICATION: DRAW ALL CONSTITUTIONAL ISOMERS OF . . .

A common homework or exam problem you may encounter begins: "Draw all constitutional isomers that have the molecular formula . . . " We devote time to working this type of problem, not merely to teach you how to tackle it, but because the ability to work these problems comfortably demonstrates an in-depth understanding of constitutional isomerism; it also provides your brain with an exercise in problem-solving skills—an absolute necessity for organic chemistry.

We jump right in with a problem in which we are asked to draw all constitutional isomers of C_4H_{10}. First realize that H atoms can form only one bond. Therefore, we cannot have a bonding arrangement of the type C—H—C. In other words, when a C atom is bonded to an H atom, the chain of atoms must be terminated in that direction. The lesson is that *the structure of a molecule is determined by the connectivity of the non-H atoms*—what is often referred to as the **backbone.** The H atoms are then placed around the outside of the molecule, where bonds are still possible, to complete the octets of those non-H atoms.

The second step is to calculate the IHD in order to determine the possible number of double bonds, triple bonds, and rings. As we saw before, this entails drawing ANY completely saturated molecule using all of the non-H atoms—in this case four C atoms—and comparing it to the molecular formula we are given. In this case, it turns out that C_4H_{10} is completely saturated. Therefore, the IHD is 0, and any molecule we come up with must contain only single bonds and no rings. Knowing this dramatically decreases the possibilities!

Given the molecular formula C_4H_{10}, the backbone must consist of only C atoms. Therefore, *each unique connectivity between just the C atoms yields a unique constitutional isomer.* One

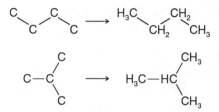

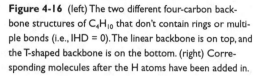

Figure 4-16 (left) The two different four-carbon backbone structures of C_4H_{10} that don't contain rings or multiple bonds (i.e., IHD = 0). The linear backbone is on top, and the T-shaped backbone is on the bottom. (right) Corresponding molecules after the H atoms have been added in.

arrangement is that in which all of the C atoms form a linear chain. That is, C^1—C^2—C^3—C^4 (the C atoms are labeled to help us keep them straight). $CH_3CH_2CH_2CH_3$ is the molecule that this corresponds to when we add in the H atoms to complete the octets on the C atoms. Another arrangement of the C atoms, possibly giving rise to a different connectivity, is that in which C^4 is disconnected from C^3, leaving C^1—C^2—C^3, and is reconnected to one of the other two Cs (either C^1 or C^2). If C^4 is connected to C^1, this yields C^4—C^1—C^2—C^3. Although the atom labels are in a different order, this connectivity is exactly the same as the original connectivity of C^1—C^2—C^3—C^4, thereby yielding the same molecule of $CH_3CH_2CH_2CH_3$. If, on the other hand, C^4 is connected to C^2, the result is a T-shaped backbone, which is unique from the linear connectivity (Figure 4-16). The T-shaped backbone is unique in that it contains a C atom that is bonded to three other C atoms—something that is not present in the linear arrangement. Filling in the remaining H atoms, the molecule becomes $HC(CH_3)_3$.

Stepping it up a notch, let's now determine all of the constitutional isomers of C_4H_9F. First realize that the IHD = 0 (convince yourself that this formula is consistent with a completely saturated molecule). Second, realize that *an F atom behaves very much like an H atom, in that it forms only one bond* (see Table 2.2). Therefore, there can be no bonding arrangement such as C—F—C. Consequently, it would be wise to add the F atom after you have constructed the various carbon backbones. Once the F is added, it can then be treated as part of the backbone, and the molecule can be completed by adding all of the Hs to complete the octets of the C atoms.

We notice that as with C_4H_{10}, there are only two different connectivities for the carbon backbone of C_4H_9F—one that is linear (C^1—C^2—C^3—C^4) and one that is T-shaped ($C(C)_3$). Working with the linear backbone, realize that there are two distinct C atoms. C^1 has exactly the same connectivity as C^4, and C^3 has exactly the same connectivity as C^2. Therefore, there are two distinct structures after adding the F—one in which the F atom is bonded to C^1 (or C^4), and one in which it is bonded to C^2 (or C^3). See Figure 4-17a and

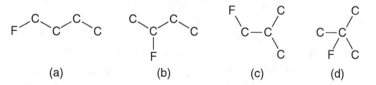

(a) (b) (c) (d)

Figure 4-17 The four different backbones of C_4F, in which there are no multiple bonds and no rings (i.e., IHD = 0).

Figure 4-17b. The final move is simply to add the nine H atoms to complete the molecule. Therefore, there are two molecules of C_4H_9F that have a linear carbon backbone.

Working with the T-shaped backbone, notice once again that there are only two distinct C atoms—the central C is the only one that is bonded to three other C atoms, and the three remaining C atoms all share the same connectivity. Therefore, when it comes time to add the F atom, there are two choices, giving rise to two distinct structures (Figure 4-17c and Figure 4-17d). Adding all of the Hs, then, yields two more distinct molecules. Reviewing, it appears that there are four total constitutional isomers of the formula C_4H_9F—two with the linear carbon backbone, and two more with the T-shaped carbon backbone.

Using a systematic methodology like this becomes more important when you attempt to draw all the constitutional isomers of $C_4H_8F_2$. We start once again with determining the IHD, which is 0. So there should be only single bonds and no rings. The various carbon backbones that can be constructed from four C atoms are the same as before—the linear one and the T-shaped one. As we just saw, within the linear carbon backbone there are two ways in which to distinctly attach the first F atom (F^1), yielding F^1—C^1—C^2—C^3—C^4 and C^1—C^2F^1—C^3—C^4 (Figure 4-17a and Figure 4-17b). We now must add the second F atom (F^2). In the structure F^1—C^1—C^2—C^3—C^4, there are now four distinct C atoms to attach it to. Even though there are four C atoms connected linearly, just as before, the addition of F^1 made C^1 different from C^4 and made C^2 different from C^3. That is, all four C atoms have different connectivities. Therefore, there are four different structures of C_4F_2 with the linear carbon backbone in which the first F atom is added to C^1 (Figures 4-18a through 4-18d).

Adding F^2 to C^1—C^2F^1—C^3—C^4 can also be done four different ways, because the addition of F^1 created different connectivities on all four C atoms (Figure 4-18e through Figure 4-18h). At first glance it appears that there are eight distinct structures of C_4F_2 with the linear carbon backbone—four with having added F^1 to C^1, and four with having added F^1 to C^2. However, in these eight structures, there are two redundancies: F^2C^1—C^2F^1—C^3—C^4 (Figure 4-18e) has exactly the same connectivity as F^1C^1—C^2F^2—C^3—C^4 (Figure 4-18b), and C^1—C^2F^1—C^3—C^4F^2 (Figure 4-18h) has the same connectivity as F^1C^1—C^2—C^3F^2—C^4 (Figure 4-18c). The reason is that, even though we created labels for the different C atoms and the different F atoms, the labels are imaginary; in reality, both F atoms are indistinguishable. In total, then, there are six distinct structures of C_4F_2 with a linear carbon backbone. Adding the H atoms, this translates into six different molecules so far.

Working now with the T-shaped carbon backbone, recall that F^1 can be added to one of two distinct C atoms to make either F^1C—$C(C)$—C or C—$CF^1(C)$—C (see Figure 4-17). In the first of these structures (Figure 4-17c), there are now three distinct C atoms to which F^2 can be added—the two terminal C atoms without F^1 have the same connectivity. That generates three different connectivities of C_4F_2 with a T-shaped carbon backbone (Figure 4-18i through Figure 4-18k). In the second of these structures (Figure 4-17d), F^2 can only

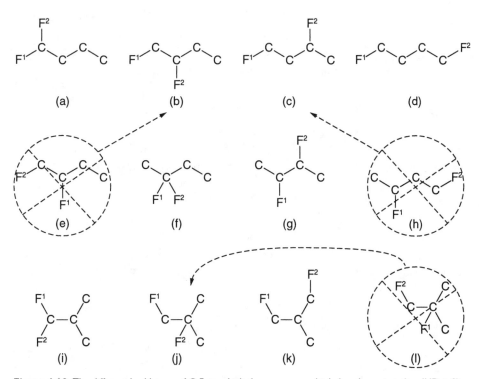

Figure 4-18 The different backbones of C_4F_2 in which there are no multiple bonds or rings (i.e., IHD = 0). The top two rows show the structures with the C atoms in a linear fashion, and the bottom row shows the structures with the C atoms forming a T-shape. Those that are crossed out are redundant to the structures indicated by the arrow.

be added to one of the terminal C atoms, but all three terminal C atoms share the same connectivity. Therefore, there is potentially one more distinct structure for C_4F_2 (Figure 4-18l). However, as before, we must be concerned with redundancy. As you can see, F^1C—$CF^2(C)$—C (Figure 4-18g) has the same connectivity as the potentially new structure, F^2C—$CF^1(C)$—C (Figure 4-18l). So the new structure is actually redundant. Consequently, there are only three total distinct structures of C_4F_2 with the T-shaped carbon backbone, which give rise to three distinct molecules. Therefore, taking into account both carbon backbones, there are nine total constitutional isomers of the formula $C_4H_8F_2$.

Let's look at one more formula—C_3H_6O. As we have been doing, it is best to use a systematic approach. First, determine the IHD. We must compare the formula we are given with a completely saturated molecule with three C atoms and one O atom. One possibility is $CH_3CH_2CH_2OH$, which has a formula C_3H_8O. Consequently, our formula has an IHD = 1, which can be either a double bond or a ring.

Before we move forward, we should identify a characteristic of the O atom that makes these types of problems more difficult than atoms such as F. The O atom can form two bonds instead of one (see Table 2-2). Therefore, if a C atom is singly bonded to an O atom, the chain does not end, because the O atom is singly bonded to another atom (in our case here that may mean a C atom or an H atom). F atoms, on the other hand, can form only one bond, so that if a C atom is bonded to an F atom, the chain is "capped off" in that direction.

Working first with a ring structure, realize that there can be three distinct rings—two rings containing three atoms (one with all Cs, and one with two Cs and the O), and one ring containing four atoms (all three Cs and the O). All three C atoms in the three-carbon ring have the same connectivity. Therefore, placing the O atom at any of those Cs gives the same structure (Figure 4-19a). The O atom has to be singly bonded to the C, because the ring structure already took care of the IHD = 1. The molecule is then completed by filling in all of the Hs. In the second ring, containing two Cs and the O, the third C atom can be bonded only to one of the ring Cs, because the O atom already is bonded twice. Those two Cs in the ring have identical connectivity; therefore bonding the third C to either of those yields identical structures (Figure 4-19b). Once the third C atom is bonded, the molecule is again completed by adding in all of the Hs. In the final ring structure—that with the four-membered ring—all of the non-Hs are already taken care of (Figure 4-19c), and so the only thing remaining is to add in the Hs.

Those isomers that do not contain a ring must contain a double bond in order to achieve an IHD = 1. The double bond could either be a C=C or a C=O double bond. And the arrangement of the atoms could be in any order. To do this systematically, let's first write out the different orders in which we can find the four non-H atoms. There are two in which the atoms are all in a row: C—C—C—O or C—C—O—C. And there is one branched arrangement, which is C—C(C)—O (a central C is bonded to two Cs and one O). Any other arrangement leads to a redundant structure.

Figure 4-19 The nine unique backbones of C_3O, with an IHD = 1.

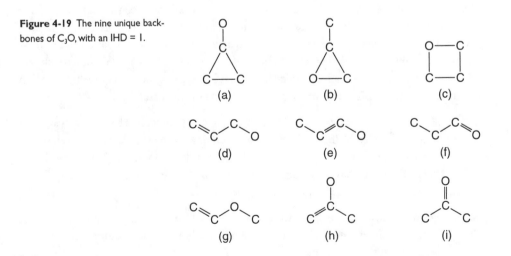

In the first arrangement of the atoms, let's now place the double bond. That gives rise to three different structures: C=C—C—O, C—C=C—O, and C—C—C=O (Figure 4-19d through Figure 4-19f). For each of these three structures, the molecule is completed by adding in the Hs. In the second arrangement (C—C—O—C), the double bond can only exist as a C=C, because the O is already bonded twice. The backbone then becomes C=C—O—C (Figure 4-19g), and adding the Hs completes the molecule.

In the branched arrangement of the atoms, the double bond can be either a C=C or a C=O, because the O is bonded to only one atom before the Hs are put in. Both noncentral Cs in that backbone have the same connectivity; therefore, placing the double bond between the central C and either of the terminal C atoms yields the same structure of C=C(C)—O (Figure 4-19h). The double bond could also have been placed between the O and the C, yielding a molecule with the backbone of C—C(C)=O (Figure 4-19i).

4.8 APPLICATION: FUNCTIONAL GROUPS

When you take your full year of organic chemistry, you will find that the chapters in your textbook are organized by functional group. A **functional group** is *a grouping of atoms having a specific connectivity and specific reactivity, which is part of a larger molecule.* Functional groups are so often employed in this course because *the reactivity of individual functional groups dictates the reactivity of the entire molecule of which they are a part.*

Table 4-1 displays several different functional groups that you will encounter in organic chemistry. Notice in the table that there are three ways in which functional groups differ. One is in the type of atoms the functional group consists of. An S—H and an O—H, for example, are different functional groups. Another way is by introducing different kinds of bonding: A C=C and a C≡C are different functional groups. And the third is by connecting together two different functional groups. A good example is a carboxylic acid functional group, which is an alcohol (O—H) group and a carbonyl (C=O) group fused together. It turns out that the reactivity of a carboxylic acid is sufficiently different from a carbonyl or alcohol that it is assigned as an independent functional group. Another example is an amide, which is a carbonyl group and an amine group fused together.

Certain types of bonding are noticeably absent in Table 4-1. C—C single bonds and C—H bonds are generally not there. That is because these arrangements of atoms are typically boring when it comes to chemical reactivity. Molecules that are comprised of only C—C and C—H bonds are called **alkanes** and tend to be *among the most unreactive molecules in organic chemistry.* In fact, alkanes, such as hexanes, are often used as organic solvents for chemical reactions and do not partake in the chemical reactions themselves.

Because C—C and C—H bonds tend not to be very interesting, entire portions of a molecule that are comprised of only these types of bonds are often ignored when it comes to

Table 4-1 Common functional groups

Functional Group	Name	Functional Group	Name
$\diagdown C = C \diagdown$	Alkene	$-\overset{\mid}{\underset{\mid}{C}}-N\diagup$	Amine
$-C\equiv C-$	Alkyne	$-C\equiv N$	Nitrile
(benzene ring)	Aromatic	$\diagdown C \diagup \overset{O}{\overset{\|}{C}} \diagdown C \diagup$	Ketone
$-\overset{\mid}{\underset{\mid}{C}}-X$ (X=F, Cl, Br, I)	Alkyl Halide	$\overset{O}{\overset{\|}{C}}{\diagdown}_H$	Aldehyde
$-\overset{\mid}{\underset{\mid}{C}}-OH$	Alcohol	$\overset{O}{\overset{\|}{C}}{\diagdown}_{OH}$	Carboxylic Acid
$-\overset{\mid}{\underset{\mid}{C}}-O-\overset{\mid}{\underset{\mid}{C}}-$	Ether	$\overset{O}{\overset{\|}{C}}{\diagdown}_N\diagup$	Amide
		$\overset{O}{\overset{\|}{C}}{\diagdown}_{OR}$	Ester

chemical reactivity. These portions of the molecule are called **alkyl substituents.** They are derived from alkanes and are represented by the letter R; think of it as standing for the "rest" of the molecule. Methanol, CH_3OH, and ethanol, CH_3CH_2OH, are types of alcohols that can both be represented by ROH. Their chemical behaviors are different, but often that difference is not important enough to necessitate drawing out the entire structure of the alkyl substituent.

Ring structures are other bonding arrangements that are noticeably absent in Table 4-1. This is because *functional groups that are part of a ring typically have very much the same chemical reactivity as functional groups that are not part of a ring.* Cyclohexanone, for example (Figure 4-20), contains a carbonyl functional group whose C atom is part of a six-membered ring. Its behavior is very similar to 3-hexanone, which contains no rings.

One exception is the **aromatic ring,** a species in which three double bonds alternate with single bonds in a six-membered ring. Its behavior turns out to be substantially different

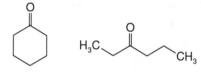

Figure 4-20 Cyclohexanone (left) has very similar chemical behavior as the hexanone molecule on the right. The carbonyl functional group being part of the ring does not dramatically change the reactivity of the functional group.

from when that triplet of double bonds is not in a ring. Understanding the reason for this requires a substantial discussion that is not appropriate for this book. Rest assured, though, that you will learn why in your full-year organic chemistry course.

Recognizing functional groups in molecules can sometimes be made difficult by the presence of rings. A great example is with acetal and hemiacetal functional groups (not shown in Table 4-1), which may or may not be parts of rings. An *acetal* group is identified by a C atom that is bonded to two alkoxy (R—O—) groups. The two alkoxy groups may be the same, such as two CH_3O groups, or they may be different, such as CH_3O and $(CH_3)_3CO$. A *hemiacetal* is identified by a C atom that is bonded to both an OH and an OR group.

Examples of an acetal and a hemiacetal functional group, not in rings, are shown in Figure 4-21a and Figure 4-21b. Figure 4-21c is a molecule of glucose, and Figure 4-21d is a dimer (two-unit structure) of glucose. In Figure 4-21c, there are several O atoms, which serve to camouflage the single hemiacetal functional group. But that functional group is easy to spot if you simply look for the C atom that is bonded to an OH group and an OR group, as previously described. There is only one hemiacetal functional group in the molecule, and that C atom is marked with an asterisk (*). This hemiacetal is often difficult to find because the R group of the OR is actually part of the ring.

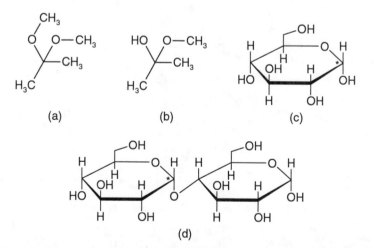

Figure 4-21 (a) Example of an acetal functional group. (b) Example of a hemiacetal functional group. (c) Molecule of glucose, containing one hemiacetal functional group, marked by an *. (d) Dimer of glucose, containing an acetal functional group, marked by an *.

In Figure 4-21d, the central C of the acetal functional group is marked with an asterisk. In this case, one OR group is part of one ring, and the R group of the other OR is part of a different ring. There is also a hemiacetal functional group in Figure 4-21d that is not marked. Can you find it?

As mentioned at the beginning of this section, functional groups dictate the chemical reactivity of entire molecules. Let's look at a few reactions that demonstrate that point. However, I must emphasize that these examples with functional groups simply illustrate the convenience of using functional groups to summarize chemical reactivity. *Organic reactions should NOT be learned in this way.* Doing so is nothing more than rote memorization, which is what we are trying to avoid. In Chapter 6 we will begin looking at reaction mechanisms, which encompass the best way to go about learning reactions.

Figure 4-22a shows that 3-hexanone reacts with NH_3 to form what is called an immine (contains a C=NH). Likewise, cyclohexanone (Figure 4-22b) reacts with NH_3 to form an

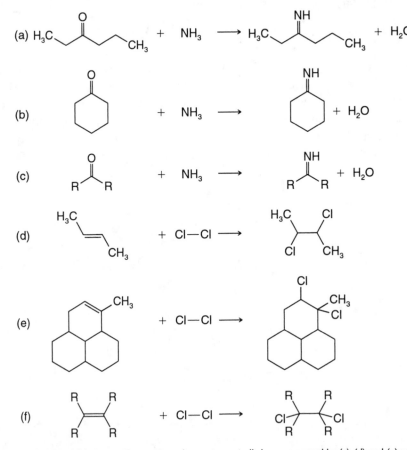

Figure 4-22 (a) and (b) are specific reactions that can generically be represented by (c). (d) and (e) are specific reactions that can generically be represented by (f).

immine. Even though the two molecules are structurally different (one is linear, and the other is cyclic), these molecules have reacted in exactly the same way, if viewed correctly. That is, in both reactions a C=O group is converted into a C=NH group (Figure 4-22c). The only difference is in the two R groups that are bonded to the C of the C=O group. In Figure 4-22a, the R groups are separate from one another. In Figure 4-22b, one R group is actually bonded to the other, forming a ring. But, as you can see, changing the nature of the R groups does not appear to have a substantial effect on the outcome of the reaction.

Another example is shown in Figure 4-22d through Figure 4-22f. In Figures 4-22d and 4-22e, two different alkenes are shown to react with Cl_2, each converting the double bond into a single bond and yielding a 1,2-dichloride—a molecule with two Cls, one on each of two adjacent C atoms. As shown in Figure 4-22f, both of these reactions can be viewed as $R_2C=CR_2$ reacting with Cl_2. In Figure 4-22d, two of those R groups are H and two are CH_3. In Figure 4-22e, one R is an H atom, one is a CH_3, and two are groups that are part of a large system that contains three rings.

4.9 APPLICATION: FORMATION OF STEREOCENTERS IN REACTIONS

In Section 4.4, we discussed how various stereoisomers (which usually contain stereocenters) react. A pair of enantiomers *must* have exactly the same chemical reactivity. A pair of diastereomers *must not* have exactly the same chemical reactivity—they may be quite different in reactivity, or they may be very similar, but there will be some difference. In this section, instead of discussing the reactivity of species that contain stereocenters, we will discuss scenarios in which stereocenters are formed.

Before we discuss this in depth, we must talk a bit more about the concept of stereocenters. Earlier, we defined a stereocenter as a tetrahedral atom that is bonded to four different groups. It is important to realize here that, for a given set of four groups attached to a given stereocenter, there are exactly two types of configurations (Figure 4-23). One is designated as the "R" configuration of the stereocenter, and the other one is the "S" configuration (we will

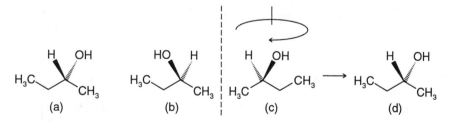

Figure 4-23 (a) R form of 2-butanol. Breaking the C—H and C—O bonds at that stereocenter and interchanging the two groups yields (b), the S form of 2-butanol. Reflecting (b) through the mirror yields (c), which, upon rotating the molecule 180° about the axis indicated, appears to be the exact same molecule as (a). Therefore, *an R stereocenter can be converted to an S stereocenter by either exchanging any two groups or by reflecting through a mirror.*

not discuss how to identify each stereocenter as R or S—that will be left to your full-year organic chemistry course). From Figure 4-23, you should clearly see that *these two configurations—R and S— are mirror images of each other* (a recurring theme in stereoisomerism). You should also see that *if you exchange any two groups of a stereocenter, R is converted to S, and vice versa*. (Note: this involves physically breaking the two bonds, rearranging the groups, and then forming two new bonds—something that does not happen spontaneously.) If you do not see this right away, use a molecular modeling kit to build the structures in Figure 4-23a and Figure 4-23b, and convince yourself that they are indeed mirror images of each other!

For the remainder of the section, we will focus on the idea that in a chemical reaction, *achirality is preserved*. Therefore, *if, in an achiral environment, a stereocenter is formed from an atom that is initially not a stereocenter, then both the R and S configurations must be formed in equal amounts*. Otherwise, an excess of one enantiomer over the other would generate a net chirality in the products. This can be better understood if we examine specific reactions in which this occurs.

2-butanone, for example, can react with a hydride anion, H⁻, to be converted ultimately into 2-butanol (Figure 4-24). Both reactant species are achiral (prove this to yourself!), and there is nothing chiral about the environment in which the reaction takes place (it would otherwise be indicated somehow). The product molecule, however, contains a single stereocenter (as we saw in Figure 4-23). Therefore, each molecule that is formed must be chiral, and the stereocenter must be either the R or the S configuration (note the R/S configuration is not explicitly shown in Figure 4-24; that is, it is not indicated which group —H or OH— is in front and which is in back). According to the previous statement, both the R and S stereocenters must be formed in equal amounts, giving rise to *equal amounts of two enantiomers*—a so-called **racemic mixture** of enantiomers.

Both R and S are formed in equal amounts because of the symmetry of the species that are reacting in the step in which the stereocenter is formed (i.e., the first step). As you can see in Figure 4-25, 2-butanone has a plane of symmetry. The H⁻ can react with 2-butanone by approaching from either side of the molecule's plane, and it does so with equal probability. As far as the H⁻ is concerned, it experiences exactly the same influence with either approach, since one side is simply the mirror image of the other.

If the approach is from the left (from the vantage point indicated in Figure 4-25), then the new H—C bond is formed on the left side, and the remaining three groups are pushed to the right; this then completes the formation of the tetrahedral C, and hence the stereocenter.

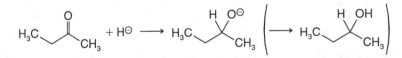

Figure 4-24 Reaction of 2-butanone with H⁻, ultimately forming molecules of 2-butanol (in parentheses). Each product molecule must be chiral, because it contains a single stereocenter. However, the R/S configuration is not indicated.

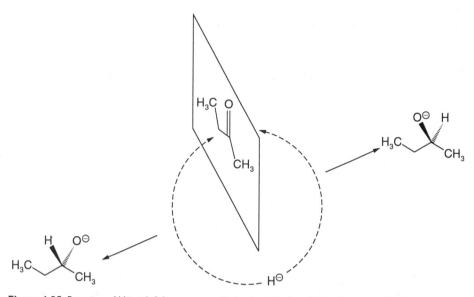

Figure 4-25 Reaction of H⁻ with 2-butanone, producing both the R and S configuration of 2-butanol. The 2-butanone molecule itself has a plane of symmetry, and the H⁻ can approach from either side, with equal likelihood. If the approach is from the left to form the H—C bond, one enantiomer is formed. If the approach is from the right, the other enantiomer is formed.

If, on the other hand, the approach is from the right, then the new C—H bond is formed on the right side, and the three remaining groups are all pushed to the left. The relationship that exists between the product molecules is that in one, an R stereocenter was formed, whereas in the other, an S stereocenter was formed. Because the approach of the H⁻ is equally likely from either side, equal amounts of both R and S configurations are formed.

Another example of a reaction that involves the generation of a stereocenter is what is called the S_N1 reaction. An example of such a reaction is that in which Br in 2-bromobutane is replaced by Cl (Figure 4-26). This reaction happens in two steps: (1) The C—Br bond is broken, liberating Br⁻, and leaving a carbocation (a species with a C⁺); and (2) reaction of the carbocation and a Cl⁻ to form the new C—Cl bond. The reactant molecule contains a single stereocenter, and therefore must be chiral. As shown in Figure 4-26, all reactant molecules may be pure R or pure S. The product molecules also contain a single stereocenter, meaning that each molecule must be chiral. However, each product molecule is formed from the intermediate carbocation species, which has a plane of symmetry (the C⁺ is trigonal planar), and is therefore achiral. As a result, the formation of the chiral product molecules, from the achiral precursor, must result in the equal formation of both R and S. In other words, *because the reaction goes through an achiral intermediate, any stereochemistry of the reactant molecules is lost, resulting in a racemic mixture of the product enantiomers.*

Finally, let's take another look at the reaction of a carbonyl (C=O) functional group with H⁻, yielding an alcohol (OH). This time, let's suppose that the species containing the carbonyl

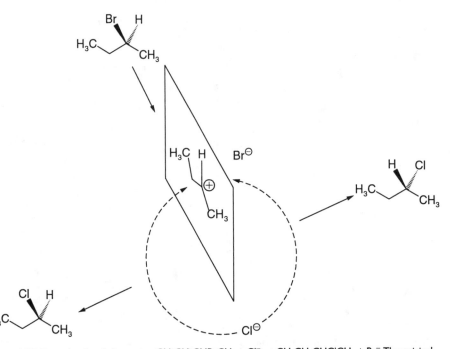

Figure 4-26 Example of an S_N1 reaction: $CH_3CH_2CHBrCH_3 + Cl^- \rightarrow CH_3CH_2CHClCH_3 + Br^-$. The original molecule is chiral and is indicated to be of one specific configuration at the stereocenter (R or S), and not a mixture. The first step of the reaction shows that the Br^- leaves, such that the stereocenter is converted from tetrahedral to planar. That generates an intermediate species that has a plane of symmetry, and when the Cl^- forms a bond to the C^+, it can approach that C^+ from either side of that plane. *Stereocenters are formed in the second step of the reaction, and because neither the carbocation nor the Cl^- are chiral, the R and S configurations must be formed in equal amounts, giving rise to a racemic mixture.*

group already contains a single stereocenter and is therefore chiral. An example is shown in Figure 4-27. Once again, the C atom of the carbonyl group is trigonal planar, meaning that the H^- can approach from either side of that plane. However, that H^- sees things differently, depending upon which side it approaches from. The reason is that the carbonyl-containing molecule is chiral, and therefore, by definition, must NOT possess a plane of symmetry. There is no plane that can be drawn such that one half is the mirror image of the other half! As a result, if the H^- experiences things differently depending upon which side it approaches from, then it will favor one approach over the other. That means that both R and S configurations are formed, but they are NOT formed in equal amounts (we cannot tell simply by looking at the structure of the molecule which of the R/S configurations will be favored, just that one will be favored over the other).

To look at this another way, the reaction is effectively taking place in a *chiral environment*. The reaction is between the hydride anion (H^-) and the C=O group. Nearby there is a stereocenter that provides the chiral environment. The stereocenter does not directly partake in the reaction, but it does influence the reaction, because it is chiral.

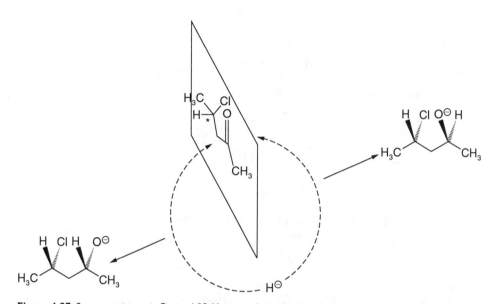

Figure 4-27 Same reaction as in Figure 4-25. However, the molecule containing the C=O has a stereocenter (marked with an *), making that molecule chiral. As a result, in the formation of the new stereocenter, both R and S configurations will be formed, but in *unequal* amounts.

Problems

4.1 Identify the specific type of relationship between each of the following pairs of molecules (that is, either *same molecules, constitutional isomers, enantiomers, diastereomers,* or *unrelated*).

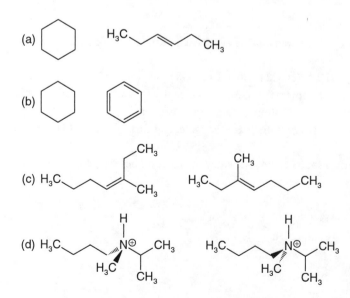

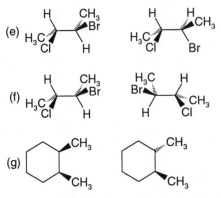

(e)

(f)

(g)

4.2 Which of the following molecules is chiral? (*Circle all correct answers*).

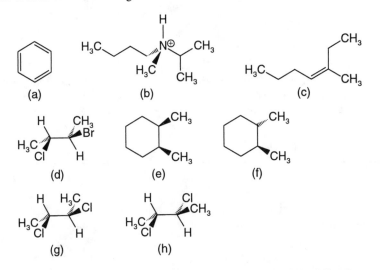

(a)

(b)

(c)

(d)

(e)

(f)

(g)

(h)

4.3 Calculate the index of hydrogen deficiency for each of the following molecular formulas.

(a) C_6H_6 (b) $C_6H_5NO_2$ (c) $C_8H_{13}NOF_2$ (d) $C_4H_{12}Si$ (e) $C_6H_5O^-$ (f) $C_4H_6O_3S$

4.4 Draw all constitutional isomers that have the molecular formula $C_3H_6OF_2$, in which the O is bonded to only one C atom. (There are 14 isomers.)

4.5 Draw all constitutional isomers that have the molecular formula C_4H_6. (There are nine isomers.)

4.6 An amide functional group can be formed by the reaction between an amine and a carboxylic acid, as shown here.

Draw the amide that is the product of the reaction involving the following molecule, in which the carboxylic acid and the amine functional groups are part of the same molecule.

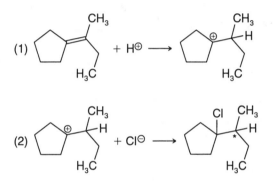

4.7 The following electrophilic addition reaction occurs in two steps, as shown. In the product, there is a single stereocenter (marked by *). Will that stereocenter be formed in equal amounts of the R and S configurations? Why or why not?

(1) [structure] + H⊕ ⟶ [structure]

(2) [structure] + Cl⊖ ⟶ [structure]

4.8 Would you expect the product of the following reaction to be formed in a racemic mixture? Explain.

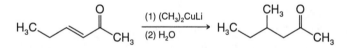

4.9 Suppose that one of the two enantiomers formed in the reaction in Problem 4.8 is reacted with H⁻, followed by reaction with water. Would the product be a racemic mixture? Why or why not?

5 Charge Stability: Charge is Bad!

5.1 INTRODUCTION

Charge is arguably the single most important factor that dictates the reactivity of a chemical species. The reason is that *opposite charges attract;* a negatively charged species has a tendency to seek out and form a bond with a positively charged species. The greater the concentration of charge—either positive or negative—on a given species, the greater is the tendency for that species to react. In other words, *charged species are inherently unstable* with respect to chemical reaction, and *the greater the concentration of charge, the more unstable that species is.* To help you remember this, we say *charge is bad,* and *more charge is worse.*

This chapter examines several different ions that you will encounter in organic chemistry. We will look at various means by which to evaluate how good (stable) or bad (unstable) the charge on that ion is, based on how well the species can *accommodate* that charge. Ultimately, you will be able to determine the relative reactivity of different species in a particular reaction, based on just their Lewis structures. And that alone is quite a powerful tool!

Of the different applications we will work with in this chapter, we spend the most time on predicting relative acid and base strengths for two reasons: (1) You are probably already somewhat familiar with acid-base (or proton transfer) reactions, and (2) those reactions are among the simplest and most common reactions that you face in organic chemistry. As we go through that application, think of the various ways in which you might be tempted to memorize—and then avoid them! For example, students often to try to memorize relative acid and base strengths of organic molecules (i.e., memorizing pK_a's) without understanding why they are what they are. As discussed in Chapter 1, doing this is yet another thing that can spell out disaster.

In addition to relative acid and base strengths, we look at relative strengths of so-called nucleophiles and the relative reaction rates of the rate determining step of the S_N1 reaction. Finally, we learn how charge plays a role in determining the best resonance contributor of a given set.

As we discuss ion stability, keep in mind that this refers to *intrinsic* stability—the stability of the species free from interactions with anything in its surroundings. Such interactions tend to complicate things. In Chapter 7, for example, we will see that interactions between ions and the solvent (i.e., "solvent effects") can dramatically affect the outcome of a reaction. Despite such complications, understanding an ion's intrinsic stability is quite effective in predicting its reactivity, even in the presence of solvent.

5.2 ATOMIC IONS

The simplest ions are atomic ions—single atoms that bear a positive or negative charge. Learning how to first determine the relative reactivity of atomic ions enables us to determine the relative reactivity of molecular ions, as discussed in the next section. To begin, let's look at the ions—both positive and negative—of some of the most common atoms in organic chemistry, including C, N, O, S, F, Cl, Br, and I.

Negative ions of the above atoms are: C^-, N^-, O^-, S^-, F^-, Cl^-, Br^-, and I^-. Each of those ions bears a negative charge, and is therefore inherently unstable with regard to chemical reaction. However, they can't accommodate that charge equally well. Differences in their stability arise from two major factors: (1) electronegativity and (2) size of the atom.

First, electronegativity. To keep things simple, let's first look only at those ions of elements in the same row of the periodic table (e.g., C, N, O, and F in the second row, and S and Cl in the third row). *The higher the electronegativity, the better the atom can accommodate negative charge.* This falls straight out of the definition of electronegativity, which is essentially a measure of how much atoms "like" electrons, or negative charge. Therefore, according to electronegativity arguments, we can say that F^- will be the most stable ion, because, of the four atoms in the first set, it is the most electronegative. Next comes O, then N, and finally C. Using the same argument, we can say that Cl^- is more stable than S^-.

If we were to use electronegativity arguments to make predictions about the relative stability of ions in the same column of the periodic table, we would actually get the wrong answer. Specifically, you know that electronegativity decreases down the periodic table. Therefore, you can argue that of F^-, Cl^-, Br^-, and I^-, F^- should be the most stable ion, then Cl^-, Br^-, and I^-. However, by experiment, it is known that the reverse order is correct. That is, I^- is the most stable of the halides, then Br^-, then Cl^-, and finally F^-.[1] Therefore, something else must be going on!

---

[1] *This order of stability is taken from gas phase proton affinities. F^- has the greatest proton affinity, and I^- has the lowest.*

That something is the different *size* of the atom on which the negative charge is isolated. Let's compare F^- and Cl^-. The Cl atom appears just below the F atom in the periodic table. This means that the valence electrons in Cl occupy a larger shell (specifically, the third shell) than the valence electrons in F (the second shell). As a result, the Cl^- ion is substantially larger than the F^- ion. Therefore, despite the fact that both ions bear the same total -1 charge, the negative charge in Cl^- has more room to spread out than it does in F^-. In other words, F^- has a more concentrated negative charge, and, again, *more charge is worse*. Therefore, we can say that Cl^- is happier, because it can better accommodate the negative charge than F^-, despite the electronegativity of F being greater.

When we compared charge stability on atoms in the same row of the periodic table, recall that we did not invoke arguments of atomic size. Although atoms in the same row certainly are different in size (size decreases from left to right across the periodic table), those differences are not as substantial as they are when adding a new shell (i.e., when moving down the periodic table).

In summary, if negative charges are on atoms from the same row, then electronegativity is what dictates the ions' relative stabilities. If the ions are in the same column, then the size of the atom that the charge is on dictates the ions' stabilities. Therefore, we can say that the stability of the halide ions, in increasing order, is: $F^- < Cl^- < Br^- < I^-$. We can also say that the stability of O^- is less than that of S^-, given that S is just below O in the periodic table.

If the charge is positive instead of negative, the same arguments can be used. Comparing the stability of C^+, N^+, O^+, and F^+, we first realize that all of these elements are from the second row, and therefore electronegativity is the deciding factor. C^+ is the most stable, because C is the *least* electronegative—it dislikes positive charge the least. Next comes N^+, then O^+, and finally F^+.

As before, the stability of atomic ions from the same column in the periodic table is dictated mostly by the size of the atom. So, among F^+, Cl^+, Br^+, and I^+, I^+ is the most stable, because it is the largest atom. I is also the least electronegative, meaning that it dislikes positive charge the least. Therefore, in the case of positive charges, both electronegativity arguments and atom size lead to the same answer.

So far we have looked at straightforward cases in which the comparisons are among atoms in the same row or in the same column of the periodic table. You may be curious if there is a way to compare the stability of ions from a different row *and* a different column; some comparisons are straightforward, but others are not. For example, if we compared the stability of O^- and I^-, we could consider the stability of a third ion, which has a row or column in common with BOTH ions—in this case, F^- is a good choice. We know that F^- is more stable than O^- because of electronegativity arguments, and we also know that I^- is more stable than F^-, because it is a much larger atom. Therefore, I^- is more stable than O^-.

On the other hand, if we compare the stability of F^- with S^-, we run into some difficulty. Let's throw in O^-, a third ion that has a row or column in common with both of those ions.

We know that F^- and S^- are BOTH more stable than O^-, the former because of electronegativity, and the latter because of size. Unfortunately, this does not help us answer the question, because we do not know which effect wins out in this case—atomic size or electronegativity. Rather than delve into it any further at this point (as doing so will likely be more confusing than enlightening), let's leave it alone and move on.

5.3 MOLECULAR IONS

There are two types of molecular ions to consider: (1) Those that have no resonance, so that the charge is isolated, or **localized,** on a single atom; and (2) those that do have resonance, which enables the charge to be distributed over at least two atoms, or **delocalized.** In this section we first consider the scenario in which the molecular ion has no resonance.

It is rather straightforward to compare two different molecular ions whose Lewis structures are quite similar, with the only major difference being the atom that formally bears the charge. For example, we can compare CH_3O^-, CH_3NH^-, $CH_3CH_2^-$, and CH_3S^-. The only significant difference among these ions is the placement of the negative charge on the O, N, C, and S atoms, respectively. In such a case, we can simply resort back to the arguments made with atomic ions. Namely, O, N, and C are in the same row of the periodic table, and because O is the most electronegative, it can accommodate the negative charge the best. Next comes N and finally C. S is in the same column of the periodic table as O, but is one row down and is therefore significantly larger. Consequently, it can accommodate the negative charge better than the O atom can. The order of stability is then: $CH_3CH_2^- < CH_3NH^- < CH_3O^- < CH_3S^-$.

Let's also consider the positively charged molecular ions $CH_3OH_2^+$, CH_3FH^+, and $CH_3SH_2^+$, where the positive charges are on the O, F, and S atoms, respectively. Because O and F are in the same row, and O is less electronegative than F, we can say that O dislikes positive charges less. Therefore, $CH_3OH_2^+$ is more stable than CH_3FH^+. Also, because S is below O in the periodic table, S is larger and can better accommodate the positive charge, which makes $CH_3SH_2^+$ more stable than $CH_3OH_2^+$. Overall, the order of stability of these three ions is: $CH_3FH^+ < CH_3OH_2^+ < CH_3SH_2^+$.

5.4 RESONANCE EFFECTS

Resonance can significantly stabilize a molecular ion. Let's look at a specific example: $CH_3CO_2^-$ and $CH_3CH_2O^-$. The Lewis structures for these ions are shown in Figure 5-1. As you can see, in both Lewis structures, the negative charge is formally on an oxygen atom. However, $CH_3CO_2^-$ has a resonance structure that appears to place the negative charge on the *other* oxygen atom. As we learned in Chapter 2, when there are multiple resonance

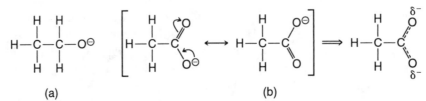

(a)   (b)

Figure 5-1 (a) Lewis structure of $CH_3CH_2O^-$. (b) Resonance structures of $CH_3CO_2^-$ are shown in brackets. The resonance hybrid is shown to the right. Note that in the resonance hybrid, there is a partial charge on each oxygen atom rather than a full charge.

structures of a given molecule, the one true structure of the species is a hybrid of those structures, which looks something like their average. Because the negative charge in $CH_3CO_2^-$ is on one O atom in one resonance structure and on the other O atom in the second resonance structure, the hybrid species has a partial negative charge on each oxygen (Figure 5-1). Compare the resonance hybrid of $CH_3CO_2^-$ to the $CH_3CH_2O^-$ ion. In the latter, there is a *full* negative charge on the oxygen atom, whereas in the former, there is a partial charge on two different oxygen atoms.

We are now in a position to apply the concept that "more charge is worse." Both ions have the same total charge of −1, but the $CH_3CH_2O^-$ ion has that negative charge *concentrated* on only one oxygen atom. That charge is spread out over two oxygen atoms in $CH_3CO_2^-$ — it is *delocalized* over the two oxygen atoms. Therefore, we can say that $CH_3CO_2^-$ is much happier, or more stable. Such an effect on the properties of a chemical species is called a **resonance effect.**

We can also examine how the stability of an ion is affected by the *number of resonance structures* it has. As an example, let's compare the $CH_3CO_2^-$ ion to the HSO_4^- ion. Each individual resonance structure of both ions places the negative charge formally on an oxygen atom. However, there are three resonance structures that can be drawn for HSO_4^- (Figure 5-2), each of which has the negative charges on different oxygen atoms. As a consequence, in the hybrid of these resonance structures, the negative charge is shared over three oxygen atoms. Only two resonance structures can be drawn for

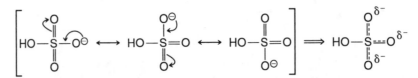

Figure 5-2 The three resonance structures of HSO_4^-. The negative charge is shared over three oxygen atoms, suggesting that each oxygen atom bears about 1/3 of a negative charge. Each oxygen atom in $CH_3CO_2^-$ (Figure 5-1), however, appears to bear about 1/2 of a negative charge.

$CH_3CO_2^-$ (Figure 5-1); the resonance hybrid shows the negative charge shared over the two oxygen atoms. Therefore, the negative charge on *each* oxygen atom in HSO_4^- is smaller than that on each oxygen atom in $CH_3CO_2^-$—specifically, one-third of a negative charge compared to one-half of a negative charge. Consequently, the HSO_4^- ion is much happier than the $CH_3CO_2^-$ ion.

5.5 INDUCTIVE EFFECTS

Inductive effects, like resonance effects, can significantly alter an ion's stability. Let's use an example to introduce the concept. Consider CH_3O^- and FCH_2O^- (Figure 5-3a). In both ions, the negative charge is formally on the oxygen atom, but the latter is significantly more stable. This is because of the F atom in place of the H atom—the only structural difference between the two ions. Furthermore, because neither ion has a resonance structure, we know that the extra stability is not due to resonance effects.

The difference in stability arises from the greater electronegativity of F than H. The F atom pulls covalently bonded electrons toward itself more strongly than does H. Consequently, the C in FCH_2O^- is more deficient of electrons than the C in CH_3O^-. In other words, the C in FCH_2O^- is hungrier for electrons than the C in CH_3O^-. Therefore, compared to the C in CH_3O^-, the C in FCH_2O^- takes more of a share of electrons from the covalent bonds it partakes in, including the C—O bond. As a result, the O atom loses some electron density, such that its negative charge is somewhat diminished. This is a good thing—since "more charge is worse," less charge must be better. The FCH_2O^- ion is therefore more stable than the CH_3O^- ion.

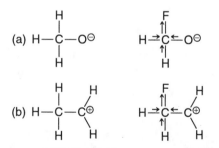

Figure 5-3 (a) (left) Lewis structure of CH_3O^-. (right) Lewis structure of FCH_2O^-. F has a higher electronegativity than H, and therefore draws electron density (negative charge) away from the C more strongly. In turn, the C in FCH_2O^- draws electron density away from the oxygen atom more strongly, which diminishes that negative charge on the O and makes the molecule happier. (b) (left) Lewis structure of $CH_3CH_2^+$. (right) Lewis structure of $FCH_2CH_2^+$. Again, the greater electronegativity of F over H draws electrons more strongly away from the neighboring C. In turn, electrons are drawn away from the C^+ more, which causes the positive charge on C^+ to increase. Because a more concentrated charge is worse, $FCH_2CH_2^+$ is less stable.

The effect that the F atom has on the stability of the ion is a result of **inductive effects.** One definition of inductive effects, therefore, is the effect that comes about from the shifting around of electron density *through covalent bonds*. In the previous inductive effect, we see that relative to H, the F atom served to *withdraw* electron density (some negative charge) from the neighboring C, which had the overall effect of withdrawing electron density from the O bearing the negative charge. We therefore say that the F atom is **inductively electron withdrawing** (compared to H), and, in such a case, **inductively stabilizes** the negative charge.

It is also possible for that same F atom to actually destabilize a molecular ion. For example, let's compare the stabilities of $CH_3CH_2^+$ and $FCH_2CH_2^+$ (Figure 5-3b). Both ions are already unhappy in that they are charged. The F atom in $FCH_2CH_2^+$, as before, pulls electron density away from the C more strongly than does the analogous H atom in $CH_3CH_2^+$. The neighboring C in $FCH_2CH_2^+$, in turn, pulls electron density away from the atoms it is bonded to, which includes the C atom bearing the full + charge. As a result, there is greater positive charge on the C^+ in $FCH_2CH_2^+$ than on the C^+ in $CH_3CH_2^+$. Consequently, $FCH_2CH_2^+$ is even more unhappy, or more unstable, than the $CH_3CH_2^+$ ion. In such a case, the withdrawing effect of F (compared to H) serves to destabilize the ion—F is **inductively destabilizing.**

Like the F atom, most substituents you encounter will be inductively electron withdrawing groups, compared to H. This is because most atoms we encounter have a greater electronegativity than H. However, there are a handful of substituents that are **inductively electron donating**—alkyl groups are one example. If we consider the simplest alkyl group, the CH_3 group, it becomes clear why this is so. The C and H atoms are very similar in electronegativity, but the C atom is slightly higher (2.5 compared to 2.2). Therefore, in each of the C—H bonds, the C gets a slightly greater share of the bonding electrons, creating an excess of electron density, and a slight δ^-, on the C atom. This buildup of negative charge, or electron density, on the C atom enables that C atom to donate the negative charge to a group that it is connected to (Figure 5-4).

Similar to the examples just presented, let's compare the stabilities of CH_3^+ and CH_3—CH_2^+, the difference being that an H in CH_3^+ is replaced by a CH_3 group. In

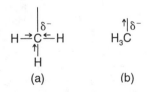

(a)   (b)

Figure 5-4 (a) Lewis structure of the CH_3 group. The C atom is more electronegative than each H atom. Consequently, there is a buildup of negative charge on the C atom. (b) The excess negative charge on the C can be partially donated through bonds, making CH_3 groups (and alkyl groups in general) electron donating groups.

$CH_3CH_2^+$, the electron-donating capability of the CH_3 group (relative to H) toward the C^+ serves to diminish some of that positive charge (some of the positive and negative charges negate one another). The smaller positive charge, as a result, is more stable.

On the other hand, alkyl groups have exactly the opposite effect on the stability of anions. A good example is the comparison of the stabilities of HO^- and CH_3O^-. In the CH_3O^- ion, the CH_3 group serves to donate negative charge toward the O atom (relative to the electron donating ability of an H atom). Because the O atom already bears a negative charge, *more charge is worse!* Therefore, the CH_3O^- ion is less stable than the HO^- ion.

Thus far, inductive effects on ion stability have been discussed in a qualitative sense; electron withdrawing groups stabilize nearby negative charges and destabilize nearby positive charges, whereas electron donating groups destabilize nearby negative charges and stabilize nearby positive charges. However, we can be somewhat quantitative as well by simply applying a few straightforward principles. All else being equal:

1   The stronger an electron withdrawing or electron donating group, the more pronounced the inductive effects.
    Example: $CH_3CHFCH_2O^-$ is more stable than $CH_3CHClCH_2O^-$.

2   The greater the number of electron withdrawing or electron donating groups, the more pronounced the inductive effects.
    Example: $CF_3CH_2^+$ is less stable than $CH_2FCH_2^+$.

3   The greater the distance between the electron withdrawing or electron donating group and the atom bearing the charge, the less pronounced the inductive effects. That is, inductive effects fall off with distance.
    Example: $CH_3CH_2CHFCH_2NH^-$ is more stable than $CH_3CHFCH_2CH_2NH^-$.

5.6   PUTTING IT ALL TOGETHER

So far, we have learned a variety of factors that dictate the stability of a species. We have learned that it makes a difference as to what kind of charge we are dealing with—positive, negative, or neutral. We have also learned that both the electronegativity and the size of the atom on which we find the charge are important. Furthermore, ion stability is affected by resonance—the greater the sharing of the charge over different atoms, the greater the stability. Finally, nearby electron withdrawing and electron donating substituents can affect ion stability by inductive effects. Given all of these factors, it can become quite overwhelming to evaluate the stability of a particular species, unless we approach such a problem systematically.

The order in which this chapter introduced the factors is essentially the order of priority for using those factors to determine relative stability. Therefore, the type of charge is generally the most important factor to consider. If one species is charged and another is not, then the charged species is typically more unstable with respect to chemical reaction. If we are comparing two ions that have the same charge, we next examine the atom(s) on which we find that charge. If there is a tie up to this point, we would further consider how much resonance serves to share that charge on different atoms. And finally, we examine differences in inductive effects. Of course, there are some situations where the order of priority for these tie breakers may differ and would lead to a different answer, but those situations are rare.

As an example of how to put together what we have learned thus far, we can determine the order of stability of the following four species:

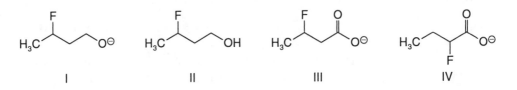

As you will see, such "ranking" problems are somewhat common in organic chemistry. How do we attack such problems, and where do we begin? *The best strategy is to make comparisons between two species at a time, and build the order as we go.* Let's start by comparing the stabilities of Structures I and II. To determine which is more stable, we simply compare their charges. Structure II has no charge, whereas Structure I has a −1 overall charge. Immediately, then, we can say that Structure I is less stable than Structure II.

Next let's bring Structure III into the picture. Because it has a −1 charge, it is clearly less stable than Structure II. But where does it rank relative to Structure I? Both have the same total −1 charge. We therefore move down the list of priorities and compare the atoms upon which we find the charge. In this case, both are on an oxygen atom. Because the tie is not yet broken, we next examine resonance. In Structure III, there is one additional resonance structure, whereas in Structure I, there is not. That additional resonance structure for Structure III is one that places the formal negative charge on the other oxygen atom, analogous to what we saw in Figure 5-1. Just as before, this delocalization of the negative charge provides extra stability, such that Structure III is more stable than Structure I. Currently, then, our order of stability is I < III < II.

Finally, we must find a place for Structure IV. Like Structure III, Structure IV has a resonance structure that allows the negative charge to be shared on both oxygen atoms, making it more stable than Structure I. However, is Structure IV more stable or less stable than Structure III? They are tied in having the same −1 charge. Moreover, they both have the same resonance structure that places the −1 charge on the other oxygen atom. We therefore must move to the last tie breaker—differences in inductive effects. It appears that all else is the same except that the F atom is closer to the −1 charge in Structure IV than in Structure III. Therefore, the effect that the F atom has on the −1 charge is more pronounced in Structure IV. Because F is electron withdrawing, that effect is stabilizing on the nearby negative charge, which makes Structure IV more stable than Structure III. Therefore, the order of stability is I < III < IV < II.

5.7 APPLICATION: STRENGTHS OF ACIDS AND BASES

Similar to the problem given in the last section, you can expect questions in organic chemistry that ask you to rank the order of acid or base strengths, given only the structures of the molecules. To do this, you could memorize the pK_a's (i.e., the numerical values that correspond to acid strength) of a dozen or more types of acids and memorize the effects that electron withdrawing and electron donating groups have on acid and base strengths (they are opposite one another). You could then say, to first approximation, that the strength of the acid of interest is approximately the same as one that is structurally similar, whose pK_a you memorized. For a better approximation, you could then apply the inductive effects you memorized—does the substituent make the acid stronger or weaker?

At this point, you should realize that this is the WRONG way to do it. The better way is to derive relative acid and base strengths from something you already know how to do—determine relative stabilities of ions. The remainder of this section guides you through doing just that.

Brönsted acids are defined as proton (H^+) donors, and bases are defined as proton acceptors. In organic chemistry, acids commonly exist as neutral molecules (like HCl) or positively charged ions (like H_3O^+). Bases commonly exist as either neutral molecules (like NH_3) or negative ions (like HO^-). We can therefore write a generic reaction for each type of acid acting as an acid (Equations 5-1 and 5-2), and we can also write a generic reaction for each type of base acting as a base (Equations 5-3 and 5-4). In Equations 5-1 and 5-2, the proton is being donated by the acid in going from reactants to products, and in Equations 5-3 and 5-4, the proton is being picked up by the base.

$$HA \rightarrow H^+ + A^- \qquad (5\text{-}1)$$

$$HA^+ \rightarrow H^+ + A \qquad (5\text{-}2)$$

$$B + H^+ \rightarrow BH^+ \qquad (5\text{-}3)$$

$$B^- + H^+ \rightarrow BH \qquad (5\text{-}4)$$

The key to identifying the stronger of two acids or two bases is to *compare the reactions* that show them acting as acids or bases—don't simply compare the acids or bases only (that is, the reactants in Equations 5-1 through 5-4). For example, if you were asked to determine which is the stronger acid, H_2O or H_2S, you must compare the following two reactions:

$$H_2O \rightarrow H^+ + HO^- \qquad (5\text{-}5)$$

$$H_2S \rightarrow H^+ + HS^- \qquad (5\text{-}6)$$

If, on the other hand, you were asked which is the stronger base, H_2N^- or $(CH_3)_2N^-$, then you must compare the following two reactions:

$$H_2N^- + H^+ \rightarrow H_3N \qquad\qquad (5\text{-}7)$$

$$(CH_3)_2N^- + H^+ \rightarrow (CH_3)_2NH \qquad\qquad (5\text{-}8)$$

If you try to solve problems such as these by comparing just the acids or bases—that is, H_2O vs. H_2S in the first problem and H_2N^- vs. $(CH_3)_2N^-$ in the second problem—then you are asking for trouble!

Once you have written out both reactions, determine which reaction has a stronger *driving force* (that is, which one is more *spontaneous,* or is more *favorable*). We then conclude that *the stronger acid (or base) is the reactant in the reaction with the stronger driving force.*

It should be evident from the last statement why we don't want to compare the acids or bases only—i.e., just the reactants. The reason is that a reaction's driving force is determined by the reactants and by the products. Sometimes differences in the driving force arise from differences in the reactants, but sometimes they arise from differences in the products! Therefore, we must compare the *whole* reaction.

In order to determine which of two reactions has the stronger driving force, let's make an analogy between a chemical reaction and a ball on the slope of a hill. A ball will spontaneously roll downhill, not uphill. That is, there is a driving force for the ball to roll downhill. And the more downhill its path, the greater the driving force. Similarly, *the more downhill a chemical process, the greater its driving force.*

How can we tell which of two chemical reactions is more downhill? It begins with the ability to associate energy with stability. Going back to our ball example, a ball on the top of a hill is less stable than that same ball at the bottom of the hill. This is because at the top of the hill, the ball possesses more energy (specifically, potential energy) than at the bottom. In much the same way, *if we are comparing two chemical species, we would say that the less stable species has more energy, and vice versa. Likewise, the more stable a chemical species, the lower its energy.*

Let's now apply what we have learned by first tackling the problem regarding which acid is stronger, H_2O or H_2S. The reactions we must compare are written out in Equations 5-5 and 5-6. Given these specific reactions, we draw a reaction *energy diagram* that illustrates how downhill one reaction is relative to the other. This energy diagram is similar to what you encountered in general chemistry, where energy is plotted on the *y*-axis (Figure 5-5). Reactants are on the left, and products are on the right.

In drawing the energy diagram, we have three tasks: (1) determine the relative energies of the reactants in each reaction, (2) determine the relative energies of the products in each reaction, and (3) connect the reactants to the products in each reaction. Sounds simple? It is if we have a good grasp of charge stability!

There is one more piece to the puzzle, though. Examining the reactions in Equations 5-5 and 5-6 once more, we notice that there are both charged species and neutral species. We

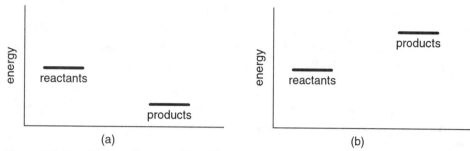

Figure 5-5 Generic energy diagrams of (a) a "downhill" reaction, and (b) an "uphill" reaction.

know how to determine the relative stabilities of two charged species. But what about two neutral species?

Fortunately, when comparing the reactants to each other and the products to each other, *we can assume that two neutral species have about the same stability;* their effects can therefore, in general, be ignored. Why is this so? It relates back to the idea that, compared to charged species, neutrals are inherently quite stable. Therefore, the *differences* in stability between two neutral species is insignificant compared to differences in stability we often see between two charged species.

Let's complete the energy diagram. On the reactant side of each equation, the only species that appear are neutral, so we place them in the energy diagram at roughly the same energy—it doesn't matter exactly where on the vertical axis (Figure 5-6). On the product side of each equation, there are two ions. Ignore the H^+ ion because it appears on the same side of the equation in both reactions—its effects cancel out. The difference in energy between the products of each reaction therefore comes down to the difference in energy between HO^- and HS^-.

From the discussions earlier in this chapter, we know that HS^- is intrinsically more stable and therefore lower in energy. We know this because HO^- and HS^- have the same negative charge, but on different atoms. Because S is below O in the periodic table, S is significantly larger in size, allowing for a much lower concentration of charge. Given its extra stability, we draw the products of Equation 5-6 with a lower energy than the products in Equation 5-5 (Figure 5-6). Again, each product energy's exact location on the vertical axis does not matter—just the energy of one product relative to the other.

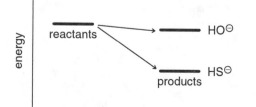

Figure 5-6 Energy diagram comparing the reactions in Equations 5-5 and 5-6. Because HS⁻ is more stable than HO⁻, the reaction in Equation 5-6 is more downhill.

Finally, we just connect the reactants of Equation 5-5 to its products and the reactants of Equation 5-6 to its products (Figure 5-6). In doing so, we see that the reaction in Equation 5-6 is more downhill, meaning that it has a greater driving force. The reactant in that equation is H_2S acting as an acid. Therefore, H_2S is a stronger acid than H_2O.

We can do exactly the same procedure to determine which is the stronger base, H_2N^- or $(CH_3)_2N^-$. To do so, we must determine which reaction is more downhill, that in Equation 5-7 or that in Equation 5-8. That requires drawing an energy diagram depicting each set of reactants going to products.

In drawing the energy diagram, we must show the relative energies of the reactants of each reaction, and we also must show the relative energies of the products of each reaction. Looking first at the reactant side of each reaction, ignore the H^+ ions, because the effect of an H^+ is exactly the same in both reactions. Don't ignore the H_2N^- or the $(CH_3)_2N^-$ species, because they are both charged and are not identical.

Both of those species have the same charge. Also, in both species, that charge appears on the same type of atom—nitrogen. In neither species are there resonance structures that can be drawn, so resonance cannot break the tie. However, inductive effects make $(CH_3)_2N^-$ *less* stable, because of the electron donation of each alkyl group to the N atom. That *increases* the negative charge concentration on N.

Because $(CH_3)_2N^-$ is less stable than H_2N^-, we know that energy of the reactants in Equation 5-8 is higher than the energy of the reactants in Equation 5-7. The products of both reactions, however, have about the same energy, because there are only neutral species that appear on that side of the equation. Figure 5-7 therefore depicts the energy diagram of the progress of each reaction, and clearly shows that the reaction in Equation 5-8 is more downhill. That means that $(CH_3)_2N^-$, which is a reactant in Equation 5-8, is a stronger base than H_2N^-.

Another problem you will frequently face in organic chemistry is whether or not a given acid-base reaction will occur. That is, if a particular acid and base were mixed, will the reaction proceed to form products? For example, we may be asked whether OH^- will react in an acid-base reaction with NH_3. *If* it were to occur, then we could write out the potential products as shown in Equation 5-9, where OH^- acts as a base, accepting a proton from H_3N, which acts as an acid.

Figure 5-7 Energy diagram comparing the reactions in Equations 5-7 and 5-8. Because H_2N^- is more stable than $(CH_3)_2N^-$, the reaction in Equation 5-8 is more downhill.

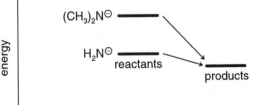

$$HO^- + H_3N \rightarrow H_2O + H_2N^- \tag{5-9}$$

In general, *an acid-base (or proton transfer) reaction such as that in Equation 5-9 will occur only if it is a downhill process—that is, if the products are more stable and therefore lower in energy than the reactants.* Proton transfer reactions do not occur to a significant extent if the process is uphill. Therefore, it becomes a matter of determining whether the products, overall, are lower in energy or higher in energy than the reactants.

In the case of the reaction in Equation 5-9, we must compare the energy of the reactants, HO^- and H_3N, to that of the products, H_2O and H_2N^-. As we have done previously, we ignore the effects of the two neutrals, assuming that they have about the same stability. The difference, then, lies with HO^- versus H_2N^-. We know that HO^- is more stable, given that O and N are in the same row of the periodic table, and that the same negative charge is on the more electronegative O atom. Therefore, the reactants are lower in energy (Figure 5-8), meaning the reaction itself is an uphill process. This reaction will not proceed as written!

Suppose, now, that we were asked whether OH^- will deprotonate CH_3SH, and, if so, what are the products? There is an added complication here. Certainly, OH^- is the base, and CH_3SH is the acid, but there is more than one possible proton that can be donated by CH_3SH—that originating from the S atom or one of the three originating from the C atom. Which is the proton that would be donated?

We must figure out which is the more acidic proton; asked in another way, is it more downhill for the S atom or the C atom to donate an H^+? To find out, compare the two complete reactions that show the species acting as acids. The reactions we must compare are those in Equations 5-10 and 5-11.

$$CH_3SH \rightarrow CH_3S^- + H^+ \tag{5-10}$$
$$HSCH_3 \rightarrow HSCH_2^- + H^+ \tag{5-11}$$

The reactants, because they are the same molecule, must be identical energy in each reaction. In the products, we can ignore the H^+. We are left with the comparison of CH_3S^- to $HSCH_2^-$. Both species have a negative charge but on different atoms. Because S is a larger atom, it can better accommodate that charge, making CH_3S^- the more stable species.

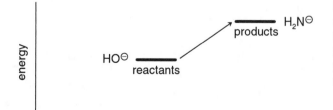

Figure 5-8 Energy diagram of the reaction in Equation 5-9. Because H_2N^- is less stable than HO^-, the proton transfer reaction is an uphill process and will therefore not proceed readily to products.

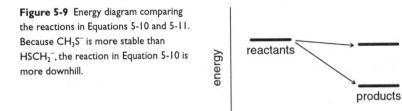

Figure 5-9 Energy diagram comparing the reactions in Equations 5-10 and 5-11. Because CH_3S^- is more stable than $HSCH_2^-$, the reaction in Equation 5-10 is more downhill.

Therefore, the reaction in Equation 5-10 is more downhill (Figure 5-9), making the proton on S the more acidic one.

Back to our original problem, we are asked whether OH^- will deprotonate CH_3SH. We must still figure out whether the proton transfer reaction, as shown in Equation 5-12, will be downhill or uphill. In comparing the energies of the reactants and products, we can, as usual, ignore the effects from the two neutrals, CH_3SH and H_2O. We are then left with determining the more stable species, CH_3S^- or HO^-. Because the S atom is larger than O, the negative charge is better accommodated in CH_3S^-. This makes the products more stable than the reactants, and, indeed, the reaction will be downhill overall (Figure 5-10). The reaction will proceed as written.

$$CH_3SH + HO^- \rightarrow CH_3S^- + H_2O \qquad (5\text{-}12)$$

5.8 APPLICATION: STRENGTHS OF NUCLEOPHILES

Reactions involving **nucleophiles** are among the most prevalent types of reactions in organic chemistry, perhaps second only to acid-base reactions. The word **nucleophile** is derived from Greek, meaning "nucleus loving" and is used to describe *any species that contains a portion that is rich in electrons;* that is, it has *an excess of negative charge.* As you will see in Chapter 6, such an excess of negative charge provides the driving force for a species to *seek out and bond to* an atom that is part of a species deficient in electrons, or one that

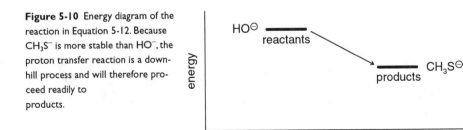

Figure 5-10 Energy diagram of the reaction in Equation 5-12. Because CH_3S^- is more stable than HO^-, the proton transfer reaction is a downhill process and will therefore proceed readily to products.

bears a partial or full positive charge. A nucleus, which is positively charged, is an example of a species that a nucleophile would be seeking—hence the name. The strength of a nucleophile, or its **nucleophilicity,** is reflected in *how fast it reacts* with a given electron-deficient species.

In general, you will encounter two types of nucleophiles: Those that are negatively charged and those that are overall neutral. Examples of negatively charged nucleophiles are HO^-, CH_3O^-, NH_2^-, Cl^-, Br^-, I^-, CN^-, $CH_3CO_2^-$, and N_3^-. Examples of nucleophiles that are overall neutral include H_2O, CH_3OH, NH_3, CH_3NH_2, and CH_3SH. All of these examples of *nucleophiles contain at least one lone pair of electrons on an atom that has excess negative charge.* The excess negative charge is what gives a nucleophile its ability to seek out an atom with excess positive charge. The lone pair of electrons is essential for the nucleophile's ability to form a bond with the electron-deficient species. That lone pair of electrons is what becomes the bond that is formed.

There is a variety of electron-deficient species that can react with a nucleophile, but we can write simple generic reactions, as shown in Equations 5-13 and 5-14, that show a nucleophile acting as a nucleophile. In Equation 5-13, the nucleophile ($Nu{:}^-$) is negatively charged, and in Equation 5-14, the nucleophile ($Nu{:}$) is neutral. In both equations, we are generically reacting the nucleophile with a carbocation, in which the C^+ is the electron-deficient species. *Notice that in the products, the existence of a formal charge depends upon whether the nucleophile is negatively charged or neutral.* If there is a formal charge in the products, notice also that it is a positive charge located on the atom directly bonded to the C.

$$R_3C^+ + Nu{:}^- \rightarrow R_3C\!-\!Nu \qquad (5\text{-}13)$$

$$R_3C^+ + Nu{:} \rightarrow R_3C\!-\!Nu^+ \qquad (5\text{-}14)$$

Throughout organic chemistry, you must know relative nucleophile strengths. There will be lists that you can memorize, but it is much more powerful to use your understanding of charge stability to derive relative nucleophile strengths. Moreover, it involves the same ideas we applied toward deriving relative acid and base strengths. Therefore, deriving the answer adds no more burden to you, contrary to the burden of having to memorize 20 additional pieces of information.

As an example, let's determine the relative nucleophile strengths of CH_3O^- and $CH_3CO_2^-$. Just as we did in comparing relative acid and base strengths, *we must compare the whole reactions* and not just the nucleophiles (the importance of this will be much clearer in our second example). The two reactions to compare are those in Equations 5-15 and 5-16.

$$CH_3O^- + R_3C^+ \rightarrow CH_3O\!-\!CR_3 \qquad (5\text{-}15)$$

$$CH_3CO_2^- + R_3C^+ \rightarrow CH_3CO_2\!-\!CR_3 \qquad (5\text{-}16)$$

Now we simply determine which reaction is more downhill. In both reactions, R_3C^+ appears as a reactant, so ignore its effects. On the product side in both reactions, only

Figure 5-11 Energy diagram comparing the reactions in Equations 5-15 and 5-16. Because $CH_3CO_2^-$ is more stable than CH_3O^-, the reaction in Equation 5-15 is more downhill.

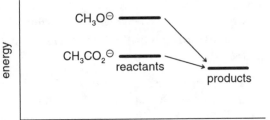

neutral species appear. Therefore, the products can be ignored. What remains is to determine the relative stability of CH_3O^- compared to $CH_3CO_2^-$. In both cases, a negative charge is on an oxygen atom, but there is resonance in $CH_3CO_2^-$ that allows the negative charge to be shared. $CH_3CO_2^-$ is therefore more stable than CH_3O^-, which makes the reaction in Equation 5-15 more downhill than that in Equation 5-16 (Figure 5-11). Consequently, the reactant in Equation 5-15, CH_3O^-, is the stronger nucleophile.

What about two neutrals? Let's determine the stronger nucleophile, H_2O or H_2S. We begin by writing the reactions that show each species acting as a nucleophile (Equations 5-17 and 5-18).

$$H_2O + R_3C^+ \rightarrow H_2O^+\!\!-\!\!CR_3 \tag{5-17}$$

$$H_2S + R_3C^+ \rightarrow H_2S^+\!\!-\!\!CR_3 \tag{5-18}$$

In drawing the energy diagram, the reactants in each equation are shown at the same energy. This is because the two neutrals, H_2O and H_2S, can be ignored, and both reactants contain the same R_3C^+ species. In the products, on the other hand, we are comparing a positive charge on an O atom to a positive charge on an S atom. Because S is significantly larger than O, it will better stabilize the charge. Therefore, the products in Equation 5-18 are lower in energy than the products in Equation 5-17 (Figure 5-12). This makes the reaction in equation 5-18 more downhill, meaning that H_2S is the stronger nucleophile.

Although we will not discuss them in this chapter, there are exceptions to the correlation between ion stability and nucleophilicity. The reason is not because of a flaw in our arguments, but rather because reactions take place in solvents, and those solvents often play a significant role. Such solvent effects will be discussed in detail in Chapter 7.

Figure 5-12 Energy diagram comparing the reactions in Equations 5-17 and 5-18. Because $H_2S^+\!\!-\!\!CR_3$ is more stable than $H_2O^+\!\!-\!\!CR_3$, the reaction in Equation 5-18 is more downhill.

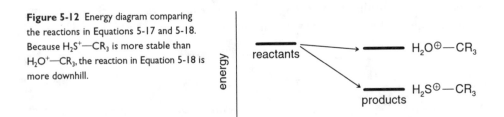

5.9 APPLICATION: THE S_N1 RATE DETERMINING STEP—RELATIVE RATES

As you will learn in the next few chapters, understanding *why* certain reaction steps have the characteristics that they do is quite powerful in predicting the outcome of reactions—ones that you have seen before and ones that you haven't. You now have the tools to understand the rate of what is called the **rate determining step** of the S_N1 reaction (see Chapter 8). This step involves only one molecule as a reactant (R—L), which breaks apart into both a positive ion and a negative ion (Equation 5-19). The positive ion is usually that of an alkyl group (R^+), which is called a **carbocation,** and the negative ion (L^-) is referred to as a **leaving group.**

$$R—L \rightarrow R^+ + L^- \tag{5-19}$$

The rate of this reaction is not particularly favorable, and, in an absolute sense, is not particularly fast. This is because the reactant is a stable neutral molecule, and both products are charged species and therefore inherently unstable. However, we are not often asked about the absolute rate of a given reaction of the kind in Equation 5-19; rather, we are usually asked to consider the relative rates of two of those reactions (i.e., which is the least slow).

Suppose we are to determine which rate determining step is faster—that involving CH_3—F or that involving CH_3—Cl. To do so, we can use exactly the same strategy that we did in working with acids and bases. We begin by writing down the two reactions to compare (Equations 5-20 and 5-21).

$$CH_3F \rightarrow CH_3^+ + F^- \tag{5-20}$$

$$CH_3Cl \rightarrow CH_3^+ + Cl^- \tag{5-21}$$

Similar to what we have seen in the previous applications in this chapter, the *faster reaction will be that which is more downhill.* Just as before, we draw an energy diagram. In doing so, we draw the reactants of each reaction at roughly the same energy, because they are both neutral. In determining the relative energies of the products, we ignore the CH_3^+ ions, given that they appear in both reactions. We are left with F^- and Cl^- to compare. Because Cl^- is more stable than F^-, we draw the products in Equation 5-21 lower in energy (Figure 5-13).

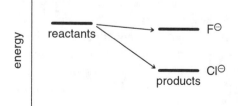

Figure 5-13 Energy diagram comparing the reactions in Equations 5-20 and 5-21. Because Cl^- is more stable than F^-, the reaction in Equation 5-21 is more downhill.

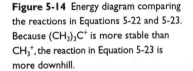

Figure 5-14 Energy diagram comparing the reactions in Equations 5-22 and 5-23. Because $(CH_3)_3C^+$ is more stable than CH_3^+, the reaction in Equation 5-23 is more downhill.

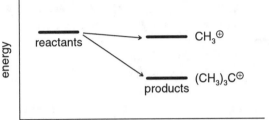

As a result, we would say that the reaction in Equation 5-21 has a greater driving force and is therefore faster than that in Equation 5-20.

In another example, we can compare the rate determining steps of reactions involving CH_3—Cl and $(CH_3)_3C$—Cl. Again, we begin by writing down the two reactions to compare (Equations 5-22 and 5-23).

$$CH_3-Cl \rightarrow CH_3^+ + Cl^- \tag{5-22}$$

$$(CH_3)_3C-Cl \rightarrow (CH_3)_3C^+ + Cl^- \tag{5-23}$$

As before, the energy diagram should show that the reactants are at roughly the same energy, given that they are both neutrals. In determining the relative energies of the products, we ignore the Cl^- because it appears in both reactions. The difference comes down to the difference in stability between CH_3^+ and $(CH_3)_3C^+$. The latter carbocation is more stable, because it has three alkyl groups surrounding the C^+, each of which stabilizes the formal positive charge by their electron-donating ability. CH_3^+, on the other hand, has no alkyl groups surrounding C^+. This leads to a greater stability for the products in Equation 5-23 than in Equation 5-22 (Figure 5-14), making the reaction in Equation 5-23 more downhill and therefore faster.

5.10 APPLICATION: THE BEST RESONANCE CONTRIBUTOR

Resonance was introduced in Chapter 2. There we mentioned that individual resonance contributors are imaginary species, and the real, true species is the resonance hybrid—an average of all of the resonance contributors of the species. However, in taking the average, not all of the resonance contributors are equal. It turns out, and should make sense, that *the resonance hybrid of a species looks most like the best, or most stable, resonance contributor.* But, given a pair of possible resonance contributors for a species, the question becomes, How do you determine which is the more stable, so that we can figure out which one the

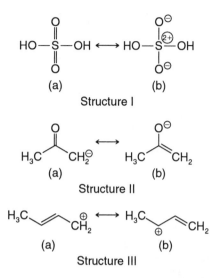

(a)   (b)

Structure I

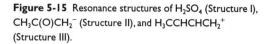

Figure 5-15 Resonance structures of H_2SO_4 (Structure I), $CH_3C(O)CH_2^-$ (Structure II), and $H_3CCHCHCH_2^+$ (Structure III).

hybrid behaves most like? Quite often, this involves the placement of a charge on one of two possible atoms.

To demonstrate this, let's determine which of the two resonance contributors is the better one in each of the examples in Figure 5-15. For Structure I, we are given two resonance contributors to compare—Structure Ia and Structure Ib. In Structure Ia, none of the atoms bears a formal charge. In Structure Ib, two O atoms bear a −1 formal charge, and the S atom bears a +2 formal charge. Knowing that "charge is bad" allows us to say that Structure Ia is the better resonance contributor, and, indeed, H_2SO_4 looks much more like that structure than Ib.

For Structure II, we are also given two resonance contributors to compare—Structure IIa and Structure IIb. Structures IIa and IIb each contain an atom that bears a −1 formal charge. In Structure IIa, that charge is on a C atom, and in Structure IIb, that charge is on an O atom. Because the two atoms are in the same row and O is more electronegative, O can better accommodate the negative charge. Therefore, Structure IIb is a better resonance contributor than IIa and, indeed, the molecular ion looks more like Structure IIb than IIa.

Finally, let's examine the two resonance contributors of Structure III that are given—Strucure IIIa and Structure IIIb. Both resonance contributors bear the same +1 charge on a C atom. In Structure IIIa, the C^+ is bonded to two H atoms and to a C that is part of the double bond. In Structure IIIb, on the other hand, the C^+ is bonded to one H, a CH_3 group, and the C that is part of the double bond. Because the CH_3 group is electron donating compared to H, the C^+ is diminished and therefore stabilized more in Structure IIIb. Therefore, Structure IIIb is the better resonance contributor.

Problems

5.1 Two resonance contributors of CH_2OH^+ are provided. It turns out that the resonance contributor whose positive charge is on the O atom is a better resonance contributor than the one whose positive charge is on the C atom, despite the fact that C is less electronegative than O. Can you explain why? (Hint: Recall from Chapter 2 what makes a viable resonance contributor.)

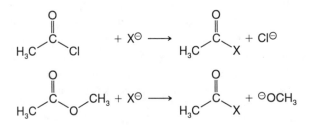

5.2 The first step in the Fischer esterification reaction, which transforms a carboxylic acid into an ester (recall the functional groups in Table 4-1), is believed to be an acid-base reaction in which the carboxylic acid is protonated (accepts an H^+). Which of the two O atoms in the carboxylic acid below is more likely to be protonated? Why?

5.3 H₃C—C(=O)—Cl and H₃C—C(=O)—O—CH₃ are both examples of "acid derivatives." Both of these reagents can undergo an "addition-elimination" reaction with a nucleophile, X^-, as shown below. If the nucleophile is the same in both reactions, which reaction would you expect to be faster? Why?

5.4 Rank the following reactions in order of increasing nucleophile strength:
  a Cl^- **b** H_2O **c** CH_3^-
  d H_2S **e** HS^- **f** I^-

5.5 Rank the following reactions in order of increasing reaction rate:

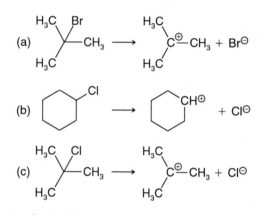

(a)

(b)

(c)

5.6 Rank the following in order of increasing base strength:

a F⁻   d CH₃NH⁻   g H₂O
b HO⁻   e NH₂⁻   h CH₃CH₂⁻
c Br⁻   f NH₃

5.7 Rank the following in order of increasing acid strength:

a CH₃CH₂CH₂CO₂H   e CH₃CH₂CCl₂CO₂H
b CH₃CH₂CHClCO₂H   f CH₃CH₂CH(CH₃)CH₂OH
c CH₃CH₂CH(CH₃)CO₂H   g CH₃CHClCH₂CO₂H
d ClCH₂CH₂CH₂CO₂H   h CH₃CH₂CF₂CO₂H

5.8 Based on arguments from this chapter, is HNO₃ a stronger or weaker acid than CH₃CO₂H? (In HNO₃, N is bonded to each O, and the H is bonded to one O.) (Hint: When the H⁺ is removed from each acid, a −1 is generated on an O atom. For which species is that −1 better stabilized?)

5.9 Draw all resonance contributors of the following ion, and rank them in order of increasing contribution to the overall resonance hybrid (Hint: There are four resonance contributors in all.)

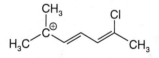

5.10 Which of the following molecules would you expect to be a stronger acid? Explain.

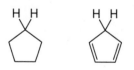

6
Reaction Mechanisms: Electron Rich to Electron Poor

6.1 INTRODUCTION

Success in organic chemistry is critically dependent on being able to make sense of reaction mechanisms. A **reaction mechanism** is a *detailed description of the individual, elementary steps that occur in an overall reaction.* An **elementary step** is *a reaction that cannot be broken down into simpler individual reactions.* An overall reaction may occur in exactly one step (i.e., A → B), in which case we would say that it occurs via a one-step mechanism. Or, it may be that the reaction occurs via a multistep mechanism, such as A → X → Y → B.

There are several reasons why it is important to be able to work comfortably with reaction mechanisms. One is that if you *understand* reaction mechanisms—that is, WHY they occur the way they do—then you will remember reactions much longer than if you try to memorize them (that is, flash-card techniques). Another reason is that if you know reaction mechanisms, and not just the reactions themselves (i.e., memorizing which reactants form which products), then you will begin to see the many similarities between reactions that otherwise seem completely different. Still another reason is that if you understand the driving forces behind the elementary steps of reactions, then you can begin to actually predict mechanisms of reactions that you have never seen before!

6.2 CURVED-ARROW NOTATION: PUSHING ELECTRONS

Understanding reaction mechanisms begins with being able to *account for all valence electrons in an elementary step of a reaction.* In other words, you must be able to trace the movement,

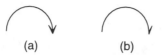

Figure 6-1 Curved-arrow notation for showing movement of electrons. (a) Double-barbed arrow indicates movement of a pair of electrons. (b) Single-barbed arrow (less common) indicates movement of a single electron.

or flow, of valence electrons. Often times this is referred to as the **pushing of electrons,** which is accomplished using curved arrows. This is important because the chemical reactivity of any species is governed almost entirely by its valence electrons, and the breaking and forming of bonds during a chemical reaction involve only valence electrons.

There are basically three rules of drawing curved arrows in reaction mechanisms. The first rule is that *curved arrows represent the movement of electrons—they do not represent the movement of atoms!* This is because the movement of electrons—the breaking and forming of bonds—describes how a reaction proceeds. The movement of atoms does not! Of course, atoms typically do move in a reaction, but this is taken care of in the curved-arrow notation by assuming that *an atom follows its own electrons* (we will see what this means shortly).

The second rule pertains to the two types of curved arrows that are used. The most common one you will encounter is the **double-barbed arrow,** which represents the *movement of a pair of electrons* (Figure 6-1a). The other type is the **single-barbed arrow,** which represents the *movement of a single electron* (Figure 6-1b). Most organic reactions you will study involve only steps in which electrons move in pairs. Most of our focus will therefore be on the use of the double-barbed arrow.

The third rule involves how to represent bonds breaking and bonds forming using the curved-arrow notation. To show **bond formation,** *the curved arrow originates from the electron(s) on an atom and points to where the bond will be.* To show **bond breaking,** *the curved arrow originates from the bond that is about to be broken and points to the atom that will acquire the electron(s).* This rule of showing bond formation and breaking applies both to σ bonds and to π bonds.

Figure 6-2a demonstrates how the curved-arrow notation is used to show the formation of a bond, where both electrons that are used to form the bond originate from the same atom—the other atom accepts a share of that pair of electrons. Note that a double-barbed arrow is used in order to represent the flow of two electrons from A. Also note that even though the arrow is used to directly show the movement of electrons, it is assumed that atom A follows those electrons so that it can get close to atom B—it is necessary for atoms A

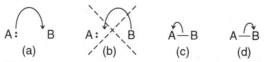

Figure 6-2 (a) Double-barbed arrow shows the movement of a pair of electrons from A, forming a covalent bond with B. (b) Incorrect way to show bond formation between A and B. (c) Double-barbed arrow shows the breaking of the bond between A and B, where the bonding pair of electrons ends up on A. (d) The bonding pair of electrons ends up on B.

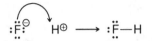

Figure 6-3 Formation of a covalent bond between F⁻ and H⁺. A lone pair of electrons from F⁻ is converted to a bonding pair of electrons. Therefore, the F atom formally loses one electron, which increases its formal charge by 1.

and B to be close to each other in order to form a bond. Finally, note the very common mistake that is shown in Figure 6-2b, which directly shows atom B moving into place to allow the bond to be formed (this breaks the first rule mentioned previously).

Figures 6-2c and 6-2d show two ways in which the bond between A and B can be broken. In both cases, both electrons from the bonding pair end up on one atom—in Figure 6-2c they end up on A, and in Figure 6-2d they end up on B. This type of bond breaking is called **heterolytic bond dissociation,** because the two electrons are not split evenly between the atoms—one atom gets both of them. Note, again, that a double-barbed arrow shows the movement of both electrons. Also, pay attention to the fact that the bond breaking is represented by the origination of the arrow from the center of the bond itself. Furthermore, that pair of electrons, which began in the form of a bond, ends up as a lone pair of electrons on the atom to which the arrow is pointing.

Before we move on, let's briefly review what happens with formal charge during the formation and breaking of covalent bonds. Recall from Chapter 2 that according to the way in which formal charge on an atom is calculated, one electron from a covalent bond is assigned to one of the atoms involved in the bond, and the other electron is assigned to the other atom. On the other hand, both electrons of a lone pair are assigned to the atom on which the lone pair is located. Therefore, if a covalent bond is formed by the donation of a lone pair of electrons from a single atom, then the formal charge increases by one on the atom that donates the lone pair (because it "formally" loses one electron), and the formal charge on the other atom decreases by one (because it "formally" gains one electron). This is exemplified in Figure 6-3, where F⁻ reacts with H⁺ to form H-F.

So far we have only discussed the formation and breaking of σ covalent bonds, but everything we have mentioned can also be applied to π bonds. The main difference is that when a π bond is broken, there is a σ bond that is still intact, allowing the two atoms to remain bonded together (Figure 6-4a). Conversely, when π bonds are formed, they form where σ bonds are already located (Figure 6-4b).

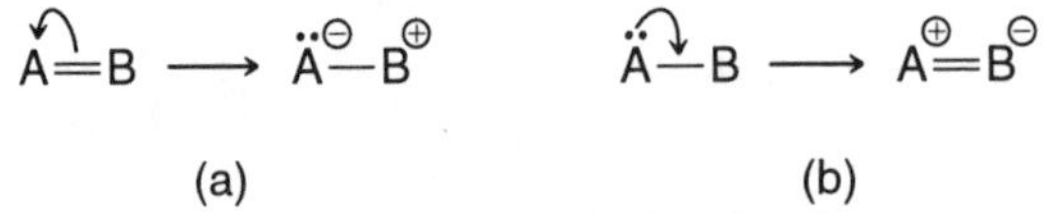

$$A=B \longrightarrow \ddot{A}^{\ominus}-B^{\oplus} \qquad A-B \longrightarrow A=B^{\ominus}$$

(a)   (b)

Figure 6-4 (a) Heterolytic breaking of a π bond, in which both of the bonding electrons end up as a lone pair on A. The A atom formally gains an electron in the process, which gives it an overall −1 charge, and the B atom formally loses one electron, giving it an overall +1 charge. (b) Formation of a π bond between A and B using a lone pair from A. Because A formally loses an electron, its charge becomes +1, and because B formally gains an electron, its formal charge becomes −1.

Figure 6-5 Formation of a bond between A⁻ and H with the simultaneous breaking of the bond between H and B. The H—B bond is forced to be broken to prevent two bonds to the H atom.

When a bond is formed on an atom in an elementary step of a reaction, the octet/duet rule may demand that a second bond on that atom be broken simultaneously. This is especially true for atoms such as H, C, N, O, and F, where the octet/duet rule strictly must not be broken. For example, H may form only one bond with other atoms. Suppose another atom donates a pair of electrons to form a bond with that H (Figure 6-5). If nothing else is done, that H would be singly bonded to two different atoms and thus would have a share of four electrons. This cannot happen. Therefore, in order for that bond to H to be formed, the original H—B bond must simultaneously be broken.

Another example is with carbon, which may form up to four bonds. If another atom donates a pair of electrons to form a covalent bond with a C that already has four bonds, then one of those four original bonds must simultaneously be broken. Figure 6-6 shows two cases in which this might occur. In Figure 6-6a, Br⁻ donates a pair of electrons to the C atom in H_3CCl, which causes the C—Cl bond to break, liberating Cl⁻. In Figure 6-6b, H⁻ donates a pair of electrons to the C atom in $H_2C=O$, causing the π bond between C and O to be broken. As a result, O gains a lone pair of electrons, and its formal charge goes from 0 to −1.

Finally, in a single elementary step of a reaction, it may be that a pair of electrons from a bond that breaks is used to form a new bond. An example of this is shown in Figure 6-7, where an HO⁻ donates a pair of electrons to form a new bond with an H atom. To avoid two bonds to H, the carbon-hydrogen bond must be broken. Instead of becoming a lone pair of electrons on the adjacent atom, those electrons from the carbon-hydrogen bond go to form a new π bond between two C atoms, causing a second σ bond to eventually be broken. The end result is the liberation of Br⁻, two atoms away from the atom that the HO⁻ directly bonds to.

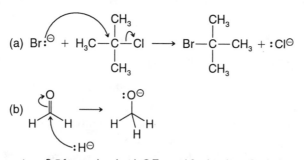

Figure 6-6 (a) Lone pair on Br⁻ forms a bond with C. To avoid five bonds to C, which would break the octet rule, Cl is forced off in the form of Cl⁻. (b) Lone pair of electrons on H⁻ forms a bond with the C atom. To avoid five bonds to C, the pair of electrons in a π bond is forced up as a lone pair on the O atom.

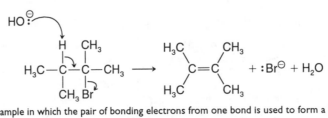

Figure 6-7 Example in which the pair of bonding electrons from one bond is used to form a bond between a different pair of atoms. The HO⁻ grabs the H, whose electrons are kicked over to form a π bond between the two C atoms, and a Br⁻ is liberated.

6.3 PREDICTING PRODUCTS FROM THE CURVED ARROWS AND DERIVING CURVED ARROWS FROM REACTANTS AND PRODUCTS

The next two sections (Sections 6.4 and 6.5) are among the most critical in this text. In Section 6.4, you will learn how to write a reasonable reaction mechanism, given an overall reaction (i.e., reactants, products, and reaction conditions). In Section 6.5, you will take this one step further and use the mechanism approach to actually PREDICT the products of reactions that you have not seen before. These two sections are important because understanding how and why reactions occur the way they do (i.e., understanding the reaction mechanism) is the key to making organic chemistry as simple as possible—as we discussed in the introduction to this chapter.

6.3a Predicting Products from Curved Arrows

In both Sections 6.4 and 6.5, we will focus on coming up with a reasonable "next step" of the mechanism by drawing reasonable electron movement—in other words, drawing reasonable curved arrows. But it is first important that you are able to draw the products of an elementary step *after* the curved arrows have been drawn in. For practice, we will go through some examples, shown in Figure 6-8. For each elementary step shown, the reactants and the curved arrows have been drawn in. We must simply write the products suggested by the curved arrows.

Before you read on, you should attempt to derive the products on your own!

In Figure 6-8a, a lone pair of electrons on the O atom in OH⁻ forms a bond with the H atom of H—OSO₃H, and the OH⁻ is assumed to follow that lone pair. At the same time, the H—OSO₃H single bond is broken, and the O atom of OSO₃H picks up a lone pair of electrons. Because that H—O single bond is broken, that H is liberated from the H—OSO₃H species. Overall, the H has been transferred from H—OSO₃H to OH⁻ (Figure 6-9a). Examining the formal charges, it appears that the O from OH⁻ should go from −1 to 0, since its lone pair of electrons is transformed into a covalent bond

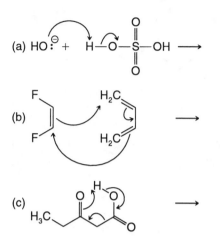

Figure 6-8 Practice problems in drawing the outcome of electron pushing.

(it formally loses one electron). On the other hand, the formal charge on the O atom in $HOSO_3H$ goes from 0 to -1, because one of its bonds becomes a lone pair (it formally gains an electron). Overall, then, the H that is being transferred does not carry with it any electrons, and therefore bears a $+$ charge—this reaction is actually an H^+ (or proton) transfer reaction, also known as an acid-base reaction.

In Figure 6-8b, it appears that six electrons are in motion simultaneously (three pairs of electrons). The π bond from the double bond of FCH=CHF is broken and subsequently forms a σ bond with a terminal C of CH_2=CHCH=CH_2. The π electrons from the double bond of the terminal C shift over to convert the C—C into a C=C. And a pair of electrons from the other π bond of H_2C=CHCH=CH_2 is converted into a σ bond back to the two-carbon species. The end result is the formation of a six-carbon ring with a C=C (Figure 6-9b). Tracing the individual pairs of electrons, notice that this occurs by the conversion of two π bonds to two σ bonds and by the conversion of one more π bond to a different π bond.

Figure 6-9 Solutions to the problems in Figure 6-8.

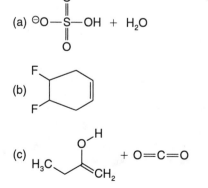

In the final example (Figure 6-8c), we again see three double-barbed curved arrows, representing the simultaneous motion of six electrons. Rather than involving two separate species, this reaction step involves only one molecule—the electrons simply shift around within that molecule. In this example, two electrons from the π bond in the C=O are converted to a σ bond between that O and the H of the O—H group. The single bond on the other side of that H is broken, and the pair of electrons is converted into a π bond between a C and an O. Finally, the C—C single bond is broken, and that pair of electrons becomes a π bond between another pair of C atoms. The net reaction is the liberation of a molecule of CO_2, and the simultaneous transfer of an H^+ from one O atom to another (Figure 6-9c).

6.3b Deriving Curved Arrows from Reactants and Products

In order to work comfortably with reaction mechanisms, we must be able to determine the flow of electrons in an elementary step of a reaction. In other words, we must be able to come up with the curved-arrow notation for an elementary step, given only the reactants and products. In Figure 6-10, we are presented with three different reactions and are asked to provide the curved arrows. To attack such a problem, compare the reactant and product species in each reaction in order to identify which pairs of electrons disappear from the reactants and which pairs of electrons appear in the products.

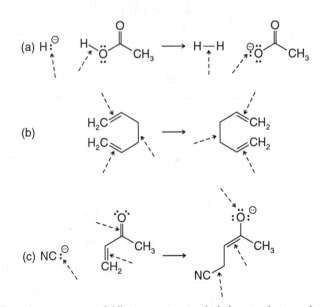

Figure 6-10 Three elementary steps of different reactions in which the curved arrows showing electron movement are not shown. The dashed straight arrows indicate electron pairs that disappear in the reactants and form in the products.

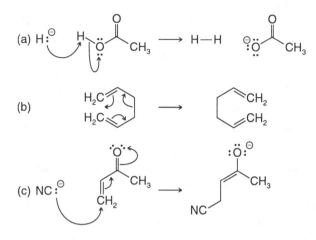

Figure 6-11 Same reactions as in Figure 6-10, with the curved arrows showing the flow of electron pairs.

Each of these pairs of electrons is indicated in Figure 6-10 by a dashed straight arrow. From there, the rest is simply drawing a curved arrow from each pair that disappears to a corresponding one that appears. Those curved arrows are provided in Figure 6-11.

6.4 DERIVING A MECHANISM: ELECTRON RICH TO ELECTRON POOR

We just saw that it is relatively straightforward to draw the products of an elementary step of a mechanism if the curved arrows are provided. We also saw that it is relatively straightforward to come up with the curved arrows for an elementary step of a reaction if we are given both the reactants and the products. But what if we are given only the reactants and are asked to come up with the curved arrows AND the products for the elementary step?

To tackle such a problem, apply the following rule: *Electrons flow from electron rich to electron poor!* This concept is important enough to repeat: *Electrons flow from electron rich to electron poor!* When drawing in the curved arrows, this means that an arrow is to originate from an electron-rich site and is to be drawn to an electron-poor site (we will elaborate on this shortly).

The reason for this rule is due to the same law of physics that we already employed in Chapter 5: opposite charges attract; like charges repel. At an electron-rich site, there is an excess of negative charge; consequently, the electrons at that site are not terribly happy. Those negatively charged electrons will therefore be repelled from the electron-rich site and will be drawn toward a positively charged site—i.e., to a site that is deficient in electrons, or electron poor.

To apply "electron rich to electron poor," let's examine the elementary steps in Figure 6-8 once more. Suppose that we were not given the curved arrows and were asked to come up with them. In Figure 6-8a, we see that the O atom of HO⁻ bears a negative charge and is therefore electron rich. The H atom of H_2SO_4 is electron poor, because the highly electronegative O atom to which it is bonded creates a bond dipole in the OH bond, leaving a δ^+ on the H atom. The curved arrow is drawn from a pair of electrons on the electron-rich O atom to the electron-poor H atom. Because that arrow represents the formation of a new bond to that H, if nothing else is done, we would have two bonds to H. To avoid this, the second arrow must be drawn in, which indicates the breaking of the OH bond on H_2SO_4.

In Figure 6-8b, *we can view a double bond as an electron-rich source*—each double bond has four electrons confined in the region between two atoms. Each C atom bonded to an F is electron poor, because the high electronegativity of F draws electrons away from C and leaves a δ^+. Electron rich to electron poor is then accomplished by drawing a curved arrow that originates from a double bond on the four-carbon species and points to one of the electron-poor C atoms on the fluorine-containing species (in Figure 6-8b, this is the curved arrow pointing from right to left). This arrow represents a new bond to the C that is bonded to F. In order to abide by the octet rule, one of the bonds already on that C has to break. This is accomplished by drawing a second curved arrow, originating from a pair of electrons of the double bond on FCH=CHF and pointing to the C atom of the other molecule. Since this second curved arrow represents the formation of a new bond to a C that already has four bonds, another bond has to break. This is the reason for the third curved arrow.

In Figure 6-8c, there are a number of electron-rich sites. One is the C=O double bond on the left—there are four electrons confined between C and O. There are also a number of electron-poor sites, one of which is the H atom of the OH bond. Electron rich to electron poor is therefore initiated by drawing a curved arrow from the C=O double bond to that H atom. To avoid two bonds to H, the OH bond must break. This is indicated by the curved arrow originating from the region between the O and H and pointing to the region where there is initially a carbon-oxygen single bond. That second arrow also represents the conversion of the carbon-oxygen single bond into a double bond. To avoid five bonds to that C, the third arrow is drawn, indicating the breaking of the carbon-carbon single bond.

6.4a Deriving a Reasonable Multistep Mechanism

Throughout a year of organic chemistry, you will encounter several reactions whose mechanisms are comprised of more than one elementary step. As we have been arguing, it is not particularly beneficial to simply memorize the reactants and products of such reactions. Rather, we must understand *how* and *why* those reactions proceed as they do. How they occur is described by the detailed mechanism of the reaction. Why they occur as they do, on the other hand, has much to do with the electron rich to electron poor concept we just learned.

To demonstrate its power, we can use electron rich to electron poor to derive a reasonable multistep mechanism, given an overall reaction. Suppose, for example, that we were asked to do so for the reaction in Equation 6-1. Be forewarned that this reaction has one of the more involved mechanisms that you will see in organic chemistry; it is comprised of nine steps. Typically it is not introduced until the second semester. We examine it at this stage only to show you that even a complex mechanism like the one you will see can be derived from the simple concepts you have learned thus far.

$$CH_3C\equiv N \xrightarrow{\;H_2O\;} CH_3CO_2H + NH_3 \qquad (6\text{-}1)$$

Before we begin, it is important to understand the physical meaning of Equation 6-1. Quite simply, it says that if we add water to CH_3CN, we will produce CH_3CO_2H and NH_3. Realize that the addition of a sample of water represents the addition of countless molecules of water (on the order of 10^{23}, or 1 mole). Therefore, although there will be a specific number of water molecules consumed for every one molecule of CH_3CN that reacts (dictated by the balanced chemical reaction), there are many water molecules that can participate behind the scenes in the elementary steps of the mechanism.

We now begin by comparing the reactants to the products in order to gain a sense of where we are starting from and where we must end up. It appears that there must be new bonds formed between C and O, and the bonds between C and N must be broken. To write a reasonable step for the formation of a C—O bond (Figure 6-12, step 1), notice that the

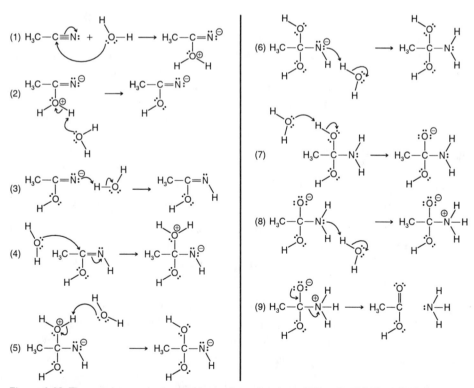

Figure 6-12 The complete mechanism of the overall reaction shown in Equation 6-1. The individual steps are described in the text.

O atom in water has two lone pairs of electrons and is also partially negatively charged (given that it is more electronegative than H). That O atom is therefore electron rich. Also, the triply bonded C bears a partial positive charge, because the N atom it is bonded to is more electronegative. Therefore, that C atom is an electron-poor site. It would then make sense to draw an arrow from a pair of electrons on the O atom to the triply bonded C atom. This then forms a bond between the O and the C. To avoid five bonds to C, a pair of electrons from the triple bond is kicked over to the N atom. Taking care of formal charges, it appears that the O atom becomes a +1, and the N atom becomes −1 (notice that the overall charge is still zero).

If we now glance at the product molecules of the overall reaction (Equation 6-1), we can gain some insight into the next couple steps. In the products, the N atom is bonded to three H atoms. After step 1, it is bonded to none. Also in the products, none of the O atoms have more than one H atom. Currently, the O atom has two H atoms. We can get closer to the overall product molecules by removing an H^+ from the O atom and adding one H^+ to the N atom. But this must be accomplished by being consistent with electron rich to electron poor. An electron-rich site is the N atom, which bears two lone pairs of electrons and a full negative charge. An electron-poor site is the O atom that bears a full positive charge.

One way to add an H^+ to the N atom and remove one from the O atom is to have the molecule do an internal proton transfer, where a lone pair of electrons on the N simply grabs an H^+ on the positively charged O. However, this is not feasible, because the O and N atoms are too far apart. A second, more reasonable way for this step to happen is for a water molecule (of which there are several) to pick up an H^+ from the positively charged O atom, and for a different water molecule to deposit an H^+ on the negatively charged N atom.

The transfer of the protons occurs in separate steps (steps 2 and 3). In step 2, the water molecule has an electron-rich oxygen atom, because it is more electronegative than the H atoms it is bonded to. That O atom therefore bears a δ^-. On the other hand, the H on the O^+ is electron poor. Therefore, we draw a curved arrow from the O on water to the H bonded to O^+. To avoid two bonds to H, we draw in the second curved arrow.

In step 3, we notice once again that the N bears a full −1 charge and is therefore electron rich. A water molecule has an electron-poor H atom, because it is bonded to a more electronegative O atom. Therefore, a curved arrow is drawn from N^- to the H on water, and the second curved arrow is drawn to avoid two bonds to H.

To get an idea of what happens next (step 4), let's again glance at the overall products in Equation 6-1. There we see that one of the products has a C atom bonded to two different O atoms, whereas after step 3 it is bonded to only one. Realizing that the first C—O bond was formed by the addition of a water molecule in step 1, let's do that again.

In order for the second water molecule to add in, step 4 looks very similar to step 1. Here again, an arrow is drawn from a lone pair of electrons on an electron-rich O atom of a water molecule and points to the electron-deficient C atom (it is still electron deficient because it is bonded to the more electronegative N and O atoms). Again, to avoid five bonds to C, a pair of electrons from the double bond is kicked over to be a lone pair on the N. The O atom of the water molecule that just added in becomes positively charged, and the N becomes negatively charged.

Step 5 is then similar to step 2, where a water molecule removes an H^+ from the positively charged O atom. Step 6 is similar to step 3, where the N^- removes an H^+ from a molecule of water. This gets us even closer to the products, because now the N atom is bonded to two H atoms instead of to one.

Realizing that in the products of the overall reaction (Equation 6-1), the N atom bears three H atoms, and there is one O atom that is not bonded to any H atoms, let's now remove another H^+ from the O atom and add one more to the N atom. This is done in steps 7 and 8, leaving a negatively charged O atom and a positively charged N atom.

Taking one last look at the overall product molecules, notice that there is a double bond between a C and an O, and the NH_3 is a separate species. This can be accomplished (step 9) by the movement of a lone pair of electrons from the electron-rich, negatively charged O atom to form the C=O. To avoid five bonds to C, and also to complete electron rich to electron poor, the C—N bond is broken. The positively charged, electron-deficient N atom gains those electrons as a lone pair, and, of course, the NH_3 molecule is liberated. The products in step 9, as you can see, are the products in the overall reaction.

Before moving on, *read this section one more time*, from beginning to end. As you go through it a second time, pay particular attention to how often we drew curved arrows from electron rich to electron poor. Also pay attention to how we continually looked at the overall reaction to help provide insight into "the next step" of the mechanism. If the next step is reasonable, then all that is left is to devise the curved-arrow notation, consistent with electron rich to electron poor.

6.5 PREDICTING PRODUCTS OF REACTIONS YOU HAVE NEVER SEEN BEFORE

The power of understanding mechanisms is truly realized when you encounter reactions you have never before seen. Even in such situations, you stand a very good chance at being able to predict the products. How? Simply by applying the rule electron rich to electron poor. We demonstrate this by examining a few reactions.

The first reaction we examine is given in Equation 6-2—the reaction of an aldehyde (CH_3CH_2CH=O) with methyl lithium (CH_3Li), followed by reaction with water. Experimentally, Equation 6-2 describes a scenario in which CH_3Li is added to a

Figure 6-13 Identification of the electron-rich (δ^-) and electron-poor (δ^+) sites for the species in the reaction in Equation 6-2. For simplicity, we treat CH_3Li as CH_3^-.

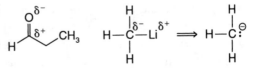

container that contains $CH_3CH_2CH=O$. After that reaction has come to a halt, water is added.

$$\xrightarrow{CH_3Li} \xrightarrow{H_2O} \quad ? \qquad (6\text{-}2)$$

To make reasonable predictions using electron rich to electron poor in the first of those two reactions, we must look more closely at each species reacting and identify the electron-rich and electron-poor sites. In the aldehyde, there is a $C=O$, in which the C atom bears a significant concentration of positive charge (Figure 6-13), because it is bonded to an electronegative oxygen atom—the C atom is therefore electron poor. For the same reason, the O atom is electron rich. The H atoms and the remaining C atoms in the aldehyde are not considered electron rich or electron poor because (1) they all have their octets/duets, (2) none of them has a lone pair of electrons, and (3) all C—H bonds are nonpolar due to the similar electronegativities of the C and H atoms.

In CH_3Li, the C atom is partially negatively charged (opposite to the situation in the aldehyde!). This is because the C atom is bonded to a Li atom, and the electronegativity of C is significantly greater than Li. As a result, the C atom is an electron-rich site, and the Li is electron poor.

Although it is technically not correct, it is easiest if we treat CH_3Li as an ionic compound, where the partial charges are full charges instead—that is, CH_3^- and Li^+. As a result, we can then *ignore* Li^+, treating it as a *spectator ion*. As we will see, *we can in general treat metal cations as spectator ions*[1].

With this in mind, we can draw a curved arrow from the negatively charged (electron-rich) C atom in CH_3^- to the electron-poor C atom of the aldehyde (Figure 6-14). The result is the formation of a C—C bond. A C atom cannot have five bonds, which forces a pair of electrons from the $C=O$ bond to be kicked up as a lone pair on the O atom. The formal charge on O then goes from 0 to -1. The reaction stops here, before the addition of water.

To determine what happens when water is added, we simply repeat the steps we did for the first reaction. First, we identify the electron-rich and electron-poor sites of the two reacting species. The product of the first reaction has a full negative charge on the O atom, making

---

[1] *As you may recall from general chemistry, metal cations like Li^+ and Na^+ are quite stable when dissolved in solution. Therefore, they typically do not deserve our attention in organic reactions and can be treated as spectator ions.*

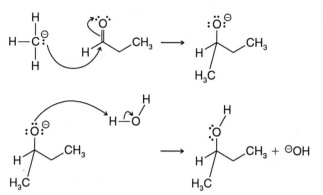

Figure 6-14 The mechanism of the reaction in Equation 6-2.

the O atom an electron-rich site in that species. In the water molecule, we note that the O atom is more electronegative than the H atoms, creating a partial negative charge on the O and leaving a partial positive charge on the Hs. Those H atoms are therefore electron poor. Next, a curved arrow is drawn from the electron-rich O atom in the product from the first reaction to the H atom in water. To avoid two bonds to that H, the other bond is broken. Overall, then, a proton (H^+) is transferred in the second reaction, producing an alcohol and OH^-.

Although we have drawn the Lewis structure of the product, we have not yet provided all the pertinent information about this reaction. Notice that in the alcohol product, there is a C atom that is bonded to four different substituents—an H atom, an OH group, a CH_2CH_3 group, and the CH_3 group that added in to form the product. Therefore, in each product molecule, that C atom is a stereocenter. Since that product molecule contains only that one stereocenter, the product must be chiral, and there are two possible enantiomers that can be formed—the R and the S. But how much of each enantiomer is formed? To answer this, we can simply apply what we learned in Chapter 4. This chiral molecule is produced from reactants that are both achiral. As a result, the chiral product must be formed in equal amounts of each enantiomer. That is, the product is a racemic mixture of the enantiomers. We are now completely finished with this reaction (*always ask yourself about the stereochemistry in a reaction!*).

Before moving on to the next example, we should explain why the C—O bond was broken in the first step of the reaction. Notice that when the CH_3^- bonds to the C of the carbonyl group, there are three possible bonds that can be broken: the C—O, the C—H, and the C—C. One reason we chose to break the C—O bond is that it is a π bond and is weaker than the other two, which are σ bonds. This can be understood if we recall that a π bond is the result of side-by-side overlap between unhybridized p orbitals. On the other hand, the σ bonds we are considering involve hybridized orbitals, which, because of the directionality of their large lobes, can result in much more of the "good" overlap we examined in Chapter 3. That additional overlap leads to a stronger bond. Therefore, of the possible bonds to break, the π bond is the weakest.

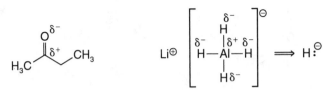

Figure 6-15 Electron-rich and electron-poor sites in the species in Equation 6-3. Because of the significant concentration of negative charge on the Hs in LiAlH$_4$, we can simplify things by treating LiAlH$_4$ as H$^-$.

The second reason we chose to break the C—O bond is due to the stability of the resulting negative charge. Depending on which of the three bonds is broken, a formal negative charge will end up on either an H, a C, or an O atom. As we have seen, if the C—O bond is broken, that negative charge ends up on the O atom, which is the most electronegative and therefore the best choice of the three.

Let's now look at another reaction. In this example, a ketone reacts with lithium aluminum hydride (LiAlH$_4$), and the product of that reaction then reacts with water (Equation 6-3). Picking the reactants apart in the same way we did in the previous example, we notice again that in the C=O bond of the ketone, the C atom is electron poor.

$$\text{(6-3)}$$

LiAlH$_4$, however, is a little trickier. It is an ionic species, comprised of two ions: Li$^+$ and (AlH$_4$)$^-$. The Lewis structure of (AlH$_4$)$^-$ is shown in Figure 6-15. Each H atom is depicted with a partial negative charge, because the electronegativity of H is greater than the electronegativity of Al (remember, the electronegativity of H, 2.2, is close to that of C, 2.5). Therefore, each H atom in that ion can be viewed as an electron-rich site.

Similar to our previous treatment of CH$_3$Li, *it is best to treat LiAlH$_4$ as a source of H$^-$ ions (or hydride ions) by simply viewing the partial negative charge on H as a full −1 charge.* The Li$^+$ and the Al^{3+} species are ignored—being metal cations, we can treat them both as spectator ions. Predicting the mechanism (Figure 6-16) then becomes much more straightforward. A curved arrow is drawn from H$^-$ to the C atom of the C=O; to avoid five bonds to C, a pair of electrons from the C=O is kicked up onto the O atom. As in the previous example, the reaction stops here. Water is then added. The electron-rich O$^-$ reacts with the electron-poor H in the water molecule, leading to a proton transfer. Once again, the product is an alcohol, which contains a newly formed stereocenter—the C atom. And, for the same reasons as in the previous example, a racemic mixture of enantiomers of this alcohol is formed.

Now is an excellent time to examine the last two examples more closely, and to demonstrate, even further, the power of learning and understanding mechanisms. When most

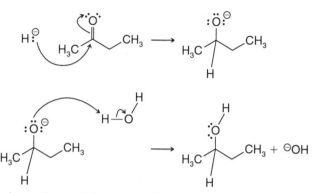

Figure 6-16 Complete mechanism of the reaction in Equation 6-3.

students look at the two reactions for which we just derived the mechanism (a carbonyl-containing molecule reacting with either LiAlH$_4$ or CH$_3$Li), they see them as different reactions. Those who attempt to learn reactions using flash cards will make different flash cards for those two reactions, and therefore must memorize them separately. However, after having seen both of their mechanisms, you should be able to see that, effectively, those two reactions are the same. *The curved arrow notation for each mechanism is identical!* The only difference is that in the first reaction, CH$_3^-$ was a reacting species, whereas in the second reaction, H$^-$ was the reacting species.

Furthermore, those who attempt to use flash cards to memorize those two reactions must memorize several different things about each reaction. In this case, they would have to memorize the reactants and products, and they would have to memorize the stereochemistry of the products—the fact that in each reaction, the product is a racemic mixture. However, *stereochemistry is something that naturally falls out of the mechanism.* If you know the steps that occur in the mechanism, and if you know how to apply what was covered in Section 4.9, then it becomes quite straightforward to predict the stereochemistry of the products. The end-of-chapter applications demonstrate this further.

To drive home the importance of learning and understanding reaction mechanisms, let's go through one more example. Equation 6-4 asks for the products when formaldehyde (H$_2$CO) is reacted with C$_6$H$_5$MgBr, followed by treatment with H$_2$O. Again, the student who relies on flash cards will make yet another flash card to try to learn to predict the products of this reaction—after all, it involves a reactant (C$_6$H$_5$MgBr) that looks quite different from the previous two reactants (LiAlH$_4$ and CH$_3$Li). However, we can predict the mechanism for this reaction just as before, and you will see that *this reaction can be viewed as no different from the previous two!*

$$ \text{(6-4)} $$

Figure 6-17 Mechanism for the reaction in Equation 6-4.

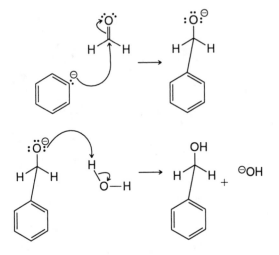

To derive the mechanism (Figure 6-17), we once again begin by identifying electron-rich and electron-poor sites. In H_2CO, the electronegativity of the O atom leaves the C atom as an electron-poor site. In C_6H_5MgBr, notice that one of the C atoms of the ring is bonded directly to the Mg atom. Because the electronegativity of C is significantly greater than that of Mg, that C atom bears a partial negative charge, making it electron rich.

At this point, you can probably see that we can treat C_6H_5MgBr as an ionic species made of $C_6H_5^-$ and $MgBr^+$, thereby making things easier for the mechanism. The negative charge is localized on a C atom of the ring (Figure 6-17), and the $MgBr^+$ species containing the metal cation can be treated as a spectator ion, as was done in the previous two reactions. We then draw a curved arrow from the electron-rich C atom of the ring to the electron-deficient C atom in H_2CO—this results in $C_6H_5CH_2O^-$. The addition of water leads to a proton transfer reaction as in the previous two mechanisms, producing $C_6H_5CH_2OH$ and OH^-.

6.6 APPLICATION: NUCLEOPHILIC SUBSTITUTION REACTIONS

Nucleophilic substitution reactions are among the first new reaction you'll encounter in organic chemistry. There are two brands of nucleophilic substitution reactions—the S_N2 (Substitution, Nucleophilic, *bi*molecular) and the S_N1 (Substitution, Nucleophilic, *uni*molecular) reactions. These reactions can be among the easiest to work with, but in actuality, students tend to find them to be among the most frustrating. Largely, this is because students tend not to focus on the mechanisms as much as is necessary. Rather, the tendency is to focus primarily on the products of the reactions, without focusing on exactly how or why it is that they were formed. In this section, as well as with the other applications in this chapter, you should focus on the hows and the whys of the reactions. In Chapter 8, we will examine the substitution and elimination reactions in their entirety. Here we will simply introduce their mechanisms.

As their names indicate, both the S_N2 and the S_N1 reactions are substitution reactions. In other words, one reacting species substitutes for some atom or group of atoms on another reacting species. The species that is doing the substituting is called a **nucleophile,** often abbreviated as Nu:⁻, or Nu:, depending upon its charge (Equation 6-5a and b, respectively). The species that is being "attacked" by the nucleophile is called the **substrate** (R—L), and the atom or group of atoms that is removed from the substrate is called the **leaving group** (L).

$$Nu:^- + R—L \rightarrow Nu—R + L^- \qquad (6\text{-}5a)$$

$$Nu: + R—L \rightarrow Nu—R^+ + L^- \qquad (6\text{-}5b)$$

Note that in both Equation 6-5a and 6-5b the total charge on the left equals the total charge on the right.

To begin to understand nucleophilic substitution reactions better, we must be able to recognize nucleophiles and leaving groups (the definitions were discussed briefly in Chapter 5). A **nucleophile** is *any species that possesses an atom with excess negative charge (either a full negative charge or a δ^-) and a lone pair of electrons.* Such a species will then "seek out" an atom that is electron deficient (i.e., its nucleus) and form a bond using that lone pair. Examples of common nucleophiles include Cl⁻, Br⁻, I⁻, HO⁻, HS⁻, RCO_2^-, C≡N⁻, H_2O, and NH_3 (the last two have no net charge).

The **leaving group** (L), on the other hand, is *an atom or group of atoms that is relatively stable (happy) upon gaining a negative charge.* You can tell that this must be true given that L⁻ is a product in Equation 6-5. Halides such as Cl, Br, and I are good examples (see Chapter 5). In general, then, when a leaving group is bonded to a C atom of the substrate, *the C—L bond is a polar covalent bond with a partial negative charge on L and a partial positive charge on the C atom. The C atom is therefore electron poor.*

Having identified electron-rich and electron-poor sites in the reactants, one way to begin the mechanism for the substitution reaction between a nucleophile and a substrate is to draw a curved arrow from the electron-rich nucleophile to the electron-poor C atom of the substrate (Figure 6-18). To avoid five bonds to the C atom, the C—L bond must be broken

Figure 6-18 Complete mechanism of the S_N2 nucleophilic substitution mechanism. The lone pair of electrons from the electron-rich nucleophile flows to the electron-poor C atom, forcing the leaving group, L, to come off as L⁻.

Figure 6-19 The complete mechanism of the S_N1 nucleophilic substitution reaction. The net reaction is similar to that of the S_N1, in that overall, Nu^- substitutes for L^-. However, this mechanism occurs in two steps. In the first step, L leaves on its own. In the second step, there is a strong driving force for the electron-rich Nu^- to form a bond with the electron-poor C^+.

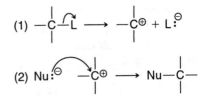

simultaneously. This turns out to be the entire mechanism for the S_N2 reaction—it occurs in just one step (note, the "2" in S_N2 does NOT indicate the number of steps in the mechanism; we discuss its meaning more fully in Chapter 8).

A second mechanism that can be drawn involves the same two arrows—a bond formation and a bond breaking—but one at a time. In the first step, L can simply leave on its own as L^-, leaving behind a carbocation, R^+ (Figure 6-19). In the second step, the electron-rich nucleophile is drawn to the electron poor R^+; the resulting formation of the Nu—C bond completes the substitution.

You might wonder why we are studying two different mechanisms for nucleophilic substitution if, as it appears, they result in the same product. The answer is that *these two mechanisms in general do not yield the same product*. The first way in which the products can differ is in their stereochemistry. If the C atom at which substitution occurs is a stereocenter, then the S_N1 mechanism will yield a mixture of stereoisomers. The reason is that in the first step of the S_N1 mechanism, the C atom bonded to L becomes trigonal planar, and is therefore achiral, in the intermediate carbocation (R^+) species. In the second step, the stereocenter is regenerated; recall from Chapter 4 that when this occurs, the stereocenter can be formed as either of the two possible configurations (R or S), resulting in a mixture of stereocenters.

An S_N2 reaction, on the other hand, does not result in a mixture of stereoisomers. Rather, only one of the two possible configurations is generated in the product. The answer as to why lies in the fact that the entire mechanism occurs in one step: the Nu—C bond is formed at the same time the C—L bond is broken. This essentially forces the Nu to attack from the side opposite L (Figure 6-20); otherwise, the Nu does not have room to come in,

Figure 6-20 The S_N2 mechanism with the L bonded to a stereocenter. Because the reaction occurs all in one step, and the product must be tetrahedral (as dictated by VSEPR, Chapter 3), the substituents (X, Y, and Z) must flip over to the other side, as indicated by the circled structure. If they don't flip over, the result is the top structure (crossed out), which is not tetrahedral.

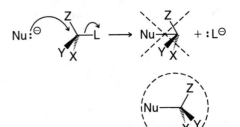

and L does not have room to leave. When this occurs, the other three substituents (X, Y, and Z in Figure 6-20) must flip over to the other side. The reason is simply that in the product, the C has four total substituents bonded to it (X, Y, Z, and Nu), meaning that it must be a tetrahedral center. If the substituents do not flip over, the result is a very oddly shaped center, which is certainly not tetrahedral. The tetrahedral geometry at the C is regained only if those substituents undergo that flip. As you can see, then, there is only one stereochemical configuration formed in the product, which is dictated by the stereochemistry at the C in the reactant substrate.

The second reason that the products of an S_N1 and an S_N2 reaction may differ is due to a potential "fork in the road" that can be taken by only one of those mechanisms. That fork in the road turns out to be available only to the S_N1 mechanism. In fact, you should quickly realize that it is impossible for such a thing to occur with the S_N2 mechanism, because everything that occurs in the mechanism (i.e., bond breaking and bond formation) happens simultaneously. In the S_N1 mechanism, however, the intermediate carbocation (R^+) has two choices: It could either react with the nucleophile to complete the S_N1 mechanism as is shown in the second step in Figure 6-19, or it could undergo an entirely different step—a **carbocation rearrangement.** You will learn more about this in your year-long organic course. We mention it here simply to make you aware of the possibilities.

6.7 APPLICATION: ELIMINATION REACTIONS

Elimination reactions are usually introduced concurrently with, or just after, nucleophilic substitution reactions. The reason is (as you will learn more about in Chapter 8) that conditions that favor S_N1 and S_N2 reactions are also favorable for elimination reactions. Equation 6-6, which outlines the generic elimination reaction, provides insight into why this might be so. You can see that, as with the previous substitution reactions, an elimination reaction involves a substrate that contains a leaving group. There are some key differences between the two reactions, however. One is that the elimination reaction involves a species acting as a base, B (i.e., an H^+ acceptor), rather than as a nucleophile. Another difference is that the reaction involves an H atom on the C adjacent to that which is bonded to L. In substitution reactions, the adjacent C was not involved. The third difference is that a double bond is formed in the product, as a result of both the H and the L being "eliminated" from the substrate.

$$B:^{\ominus} + \underset{\underset{L}{|}}{\overset{\overset{H}{|}}{-}}\overset{|}{\underset{|}{C}}-\overset{|}{\underset{|}{C}}- \longrightarrow -\overset{|}{C}=\overset{|}{C}- + BH + :L^{\ominus} \qquad (6\text{-}6)$$

Similar to nucleophilic substitution reactions, there are two different mechanisms for elimination—the E1 and the E2 mechanisms. The E1 mechanism takes place in two steps, whereas the E2 mechanism occurs in one step (again, the "1" and "2" do NOT correspond

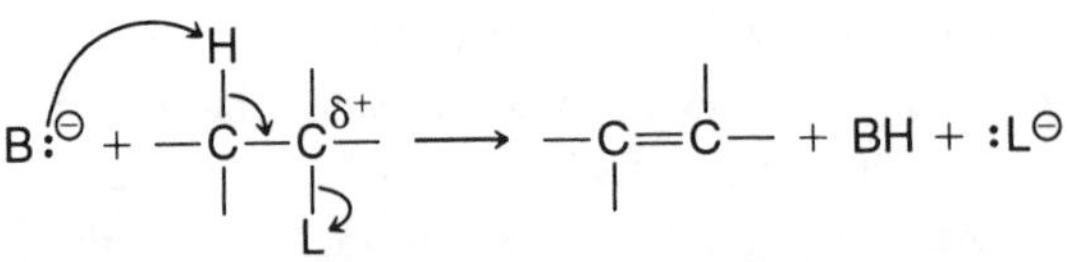

$$B:^{\ominus} + \ -\overset{|}{\underset{|}{C}}\overset{H}{\overset{|}{-}}\overset{\delta^+}{\underset{\underset{L}{\overset{|}{\downarrow}}}{C}} \longrightarrow \ -\overset{|}{C}=\overset{|}{C}- \ + \ BH \ + \ :L^{\ominus}$$

Figure 6-21 Complete mechanism for the E2 elimination reaction. Electrons flow from the electron-rich B⁻ to the H atom adjacent to the C with L. The pair of electrons from the C—H bond migrate over to form a second bond between that C atom and the neighboring C atom that is electron poor. The leaving group is kicked off in order to avoid five bonds to C.

to the number of steps in the mechanism). In the E2 mechanism (Figure 6-21), a curved arrow is drawn from the electron-rich site in the base to the H atom adjacent to the C with the L. The driving force for this is the fact that the C containing the L, in general, has a δ^+ and is therefore electron poor. When the electron-rich base forms a bond with the H, the C—H bond must be broken, and those electrons migrate over to form a bond with the electron deficient C atom (hence, electron rich to electron poor). Finally, to avoid five bonds to C, the leaving group must leave.

In an E1 mechanism (Figure 6-22, which shows it occurring in two steps), the first step involves the leaving group spontaneously leaving (exactly the same as the first step in the S_N1 mechanism). The result is the formation of a negatively charged L⁻ and a highly electron-poor carbocation (R^+). In the second step, a pair of electrons from the electron-rich base then forms a bond with the H adjacent to the C^+; this liberates the bonding pair of electrons from the C—H, which then moves over to form a second bond with the C bearing the positive charge. Overall, then, we have once again electron rich to electron poor.

Although the E1 and E2 mechanisms both result in the formation of a C=C via the elimination of an H⁺ and an L⁻, there are two main differences in their products—similar to the differences between the S_N1 and S_N2 products. The first is the stereochemistry. In general, the E2 mechanism will produce either the cis or the trans isomer (but not both), whereas the E1 mechanism will produce a mixture of both of those diastereomers (the reasons why will be left to your full-year course). The other difference is in possible side reactions. Because the E2 mechanism occurs all in one step, only the E2 product is formed. On the

Figure 6-22 Complete mechanism of the E1 reaction. As opposed to the E2 mechanism, the E1 occurs in two steps. In the first step, the leaving group leaves on its own (exactly the same as the first step of the S_N1 mechanism). In the second step, electrons from the electron-rich base flow to the H atom, and the bonding pair of electrons is kicked over to form a second bond on the electron-poor C atom.

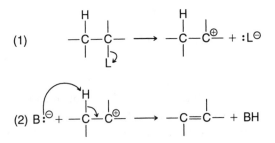

other hand, because the E1 mechanism occurs in two steps, there may be reactions involving the intermediate carbocation, which compete with the second step of the E1 mechanism. Carbocation rearrangement reactions, as mentioned before in connection with the S_N1 mechanism, are common side reactions.

6.8 APPLICATION: ELECTROPHILIC ADDITION REACTIONS

Electrophilic addition is viewed as the reverse of elimination reactions. A molecule of H—L *adds across the* C=C *double bond*. An H^+ adds to one of the C atoms, an L^- adds to the other C atom, and the double bond is converted into a single bond. Once again, the mechanism (Figure 6-23) can be derived by applying electron rich to electron poor.

In the first step of the mechanism, the H—L reacts with the C=C. In the molecule of H—L, the higher electronegativity of L induces a partial positive charge on the H, so that the H is electron poor. In the alkene, the C=C can be viewed as electron rich, given that there are four electrons confined between two C atoms. Electron rich to electron poor is therefore from a pair of electrons from the C=C to the H of H—L. To avoid two bonds to H, the H—L bond is broken, producing an intermediate carbocation and L^-. In the second step of the mechanism, the L^- is electron rich, and the C^+ is electron poor. Electrons flow from the L^- to the C^+, thereby forming the C—L bond and completing the reaction.

Notice that in the product of the electrophilic addition reaction, there are two trigonal planar C atoms that are converted to tetrahedral geometries. Therefore, there is the *potential*

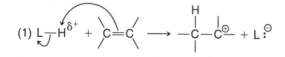

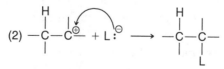

Figure 6-23 Complete mechanism of the electrophilic addition reaction, in which a molecule of HL adds across a C=C double bond. In the first step, a pair of electrons from the electron-rich double bond flows toward the electron-poor H, forcing the bond to be broken on the other side. In the second step, electrons from the electron-rich L^- flow toward the electron-poor C^+.

of creating new stereocenters; we must therefore consider the stereochemistry of this reaction. A specific example is shown in Equation 6-7. In the product, there is one stereocenter—the C atom that contains an H atom, a Br atom, a CH_3 group, and a CH_2CH_3 group. Therefore, each product molecule is chiral. Applying what we learned from Chapter 4, we realize that none of the reactants is chiral. Consequently, the product must be a racemic mixture of enantiomers.

$$CH_3CH=CHCH_3 + H-Br \rightarrow CH_3CHBrCH_2CH_3 \qquad (6\text{-}7)$$

Finally, we must recognize that when a molecule of H—L adds across a double bond, it can do so in two ways. The H^+ can add to either of the two C atoms of the C=C, and the L^- adds to the remaining C. If the two C atoms of the C=C are indistinguishable (that is, if the alkene molecule is symmetric), then only one product is formed (see Equation 6-7). If, however, those C atoms are different, then there will be two potential products that can be formed. In general, there is a preference for the formation of one of those products over the other. Such *a preference for reaction at one site (in this case, a C atom) over another* is called **regiochemistry.**

The mechanism for electrophilic addition is used to determine the regiochemistry of the reaction—that is, which of the two possible products is preferred. From Figure 6-23, realize that once the H^+ adds to one C atom in the first step of the reaction, the position where the L^- adds in has already been determined—it must be the other C atom of the double bond. Therefore, *the regiochemistry of this reaction is dictated by which C atom the H^+ wants to add to.* To figure out which C atom the H^+ prefers, look at the different carbocation products that would be formed. *The carbocation that is the most stable will be formed in the greatest amount* (this is one form of what is called **Markovnikov's Rule**).

In determining the most stable carbocation, let's work with a specific example of an asymmetric alkene, such as $CH_3CH=CH_2$. Protonation of one of the C atoms of the double bond (the first step of the mechanism) leads to either $CH_3CH_2CH_2^+$ (if the H^+ adds to the central C atom) or $CH_3CH^+CH_3$ (if the H^+ adds to the terminal C atom). In the first of those two carbocation species, the C atom bearing the positive charge is bonded to only *one alkyl group* (i.e., CH_3CH_2)—it is a **primary** carbocation. In the second carbocation, it is bonded to *two alkyl groups* (i.e., two CH_3 groups)—it is a **secondary** carbocation.

We learned in Chapter 5 that alkyl groups are inductively electron-donating species. They donate some electron density (concentration of negative charge) through the σ bonds to the formal positive charge on the C atom, which is a stabilizing effect. The greater the number of alkyl groups, the more stable the carbocation. Therefore, the second carbocation is the more stable of the two (i.e., the secondary carbocation) and will be preferred in the reaction. With that carbocation, the addition of L^- (i.e., the second step of the mechanism) yields CH_3CHLCH_3. This product, therefore, is the one that is formed in the greatest amount in the overall reaction. It is the major product.

Problems

6.1 Show the curved-arrow notation for each of the following reactions, in which both the reactants and products are given. Draw each structure with enough detail that shows all of the pertinent electron pairs.

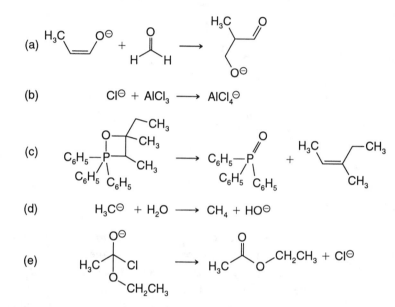

(a)

(b) $Cl^{\ominus} + AlCl_3 \longrightarrow AlCl_4^{\ominus}$

(c)

(d) $H_3C^{\ominus} + H_2O \longrightarrow CH_4 + HO^{\ominus}$

(e)

6.2 Write the products of the following reactions, for which the curved arrows have been provided. Include all nonzero formal charges on the atoms.

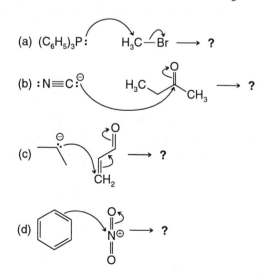

(a) $(C_6H_5)_3P\colon$ $H_3C\text{—}Br \longrightarrow$?

(b) $\colon N\!\!\equiv\!\!C\colon^{\ominus}$ $H_3C\diagdown$ $CH_3 \longrightarrow$?

(c) $\longrightarrow$?

(d) $\longrightarrow$?

6.3 Predict the products of each of the following reactions. (Hint: Draw the structure of each reactant with as much detail as is necessary, and identify *electron rich* and *electron poor*. Draw curved arrows from electron rich to electron poor, and provide the products suggested by the arrows.)

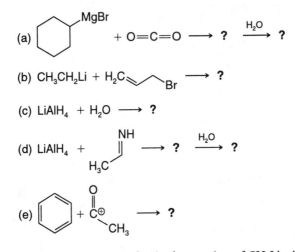

(a)

(b) $CH_3CH_2Li + H_2C$⟍⟍Br ⟶ **?**

(c) $LiAlH_4 + H_2O$ ⟶ **?**

(d) $LiAlH_4 +$ ⟶ **?** $\xrightarrow{H_2O}$ **?**

(e)

6.4 In Figure 6-14, we saw that in the reaction of CH_3Li with a compound containing a carbonyl group (i.e., C=O), the CH_3^- species bonds to the C of the C=O. If, however, the C=O is part of a carboxylic acid functional group (i.e., RCO_2H), a different reaction takes place, despite the fact that a carboxylic acid functional group also conatins a C=O. Using electron rich to electron poor, write the curved-arrow notation for the reaction of CH_3^- with RCO_2H, and predict the products of that reaction.

6.5 Write the detailed mechanism (including curved arrows) for the following reaction. Using the mechanism, predict the products, including stereochemistry if appropriate.

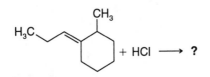

+ HCl ⟶ **?**

6.6 Draw the complete mechanism for the following reactions. Using the mechanism, predict the products, including stereochemistry if appropriate. (Note: D is deuterium, which is H with an additional neutron; its chemical reactivity can be treated identically to that of H.)

+ $LiAlH_4$ ⟶ **?** $\xrightarrow{D_2O}$ **?**

(Note: This reaction is a twist on one that you have seen before—instead of H_2O as the second reagent, D_2O [deuterated water, or deuterium "labeled" water] is used. You should know that a problem similar to this one has appeared on the MCAT exam. You probably won't learn this exact reaction in your organic class. Therefore, this reaction forces you to extend your knowledge of mechanisms. Those who rely primarily on flash cards will not know how to answer this question, because it appears to be an entirely new reaction—one they did not have a flash card for . . . .)

6.7 Predict the products of the following reaction (see Problem 6.6):

$$+ \; LiAlD_4 \longrightarrow ? \; \xrightarrow{H_2O} \; ?$$

6.8 ^{18}O denotes an oxygen atom that has 8 protons and 10 neutrons (for a total of 18 amu) It therefore has 2 neutrons more than the usual oxygen atom that has 8. If such an atom is found as the oxygen atom in water, the result is $^{18}OH_2$, called oxygen-18 labeled water. Suppose that oxygen-18 labeled water is used in the second step of a reduction of a ketone using $LiAlH_4$, as shown in the following reaction. Using the mechanism that we learned for this reaction, predict the products, including stereochemistry if appropriate.

$$\xrightarrow{LiAlH_4} \; \xrightarrow{^{18}OH_2} \; ?$$

6.9 Predict the product of the following reaction. (Hint: In the first step, to determine how $N\equiv C^-$ will react with the carbonyl, recall the mechanisms in which H^- and R^- react with a carbonyl. For the second step, water can act as an acid to protonate the product of the first step. It will subsequently facilitate the mechanism shown in Figure 6-12.)

$$\xrightarrow{NC^-} \; \xrightarrow{H_2O} \; ?$$

7 Intermolecular Interactions

7.1 INTRODUCTION

As you saw in Chapter 5, charge is integral in the stability of organic molecules and ions. In Chapter 6, you saw how the distribution of charge within a molecule is important in a reaction between two molecules—that is, where the electron-rich and the electron-poor sites are. In this chapter, you will see that charge, and distribution of charge, dictate the type and the strength of the interaction that exists *between* two separate molecules—either two of the same or two different molecules. In other words, these **intermolecular interactions** (or **intermolecular forces**) dictate how well two molecules "stick together"—or, how much energy it would cost to pull the two molecules apart.

The important intermolecular interactions we will examine are (1) ion-ion interactions, (2) ion-dipole interactions, (3) dipole-dipole interactions, (4) hydrogen bonding, (5) dipole–induced dipole interactions, and (6) induced dipole–induced dipole interactions (also called dispersion forces, or Van der Waal's forces). The names of these interactions are descriptive of the species that are involved. For example, ion-ion interactions are what keep two ions stuck together; dipole–induced dipole interactions are those that keep a neutral molecule that has a net molecular dipole stuck to one that has an "induced dipole moment" (which we will introduce later). It is important to keep in mind, however, that even though these interactions involve different types of species, all intermolecular interactions originate from the same physical phenomenon: *Opposite charges attract, and the stronger the concentration of one or both of those charges, the stronger the two species stick together.*

The applications at the end of this chapter demonstrate the importance of intermolecular interactions in dictating certain physical properties (e.g., boiling point, melting point, solubility) of organic molecules, as well as the dramatic effects that solvents can have on the outcome of an organic reaction (i.e., solvent effects). It is therefore crucial that you understand the relative strengths of the various intermolecular interactions, so you can predict the effects they have in different situations.

To do that, you must first be able to identify the type of intermolecular interactions that are present between two molecules, given only their Lewis structures (similar to predicting relative stabilities of molecular species given only their Lewis structures, as we did in Chapter 5). Therefore, this chapter will bring together many of the concepts that we covered in Chapters 2 and 3—the ones that heavily involve molecular structure. In particular, we will revisit three-dimensional molecular geometry (VSEPR theory), net molecular dipole, and formal charge.

Even though this chapter takes you through the origin of these intermolecular interactions, most students want to simply memorize the names of the interactions and their relative strengths of interaction. It is very important that you do NOT do this! If you try to memorize various aspects of intermolecular interactions, along with the numerous other bits and pieces throughout the course, then intermolecular interactions will be among the earliest to fall by the wayside. But intermolecular interactions permeate organic chemistry, and you will likely revisit them on your final exam or on the MCAT—either directly or (worse) indirectly.

7.2 ION-ION INTERACTIONS

Ion-ion interactions are conceptually the simplest of the intermolecular interactions and therefore do not require a substantial discussion. As the name suggests, these interactions exist between *two ions*. So, for example, a positive ion like Na^+ is attracted to a negative ion like Cl^-. In fact, as you are already aware, strong ionic bonds are formed between those two ions, making a very stable ionic compound—NaCl, or table salt.

Ion-ion interactions are the strongest that you will encounter in organic chemistry, evidenced by the extremely high melting points and boiling points we often see with ionic compounds (see Section 7.9). This is consistent with the earlier statement that the stronger the concentration of charge, the stronger the interaction. Ions bear full charges—positive or negative—and therefore have the strongest concentration of charge of any organic chemistry species.

7.3 ION-DIPOLE INTERACTIONS

As the name suggests, ion-dipole interactions are the attractive interactions that exist between an ion—either positive or negative—and a neutral molecule that has a net molecular dipole. One example is the interaction between an Na^+ ion and a molecule of water. Another is between a molecule of water and a Cl^- ion. As you may have guessed, and as you will see at the end of this chapter, these are the interactions that are primarily responsible for water's ability to dissolve NaCl.

Figure 7-1 (a) Shows the attraction between the positive charge of Na⁺ and the partial negative charge of the dipole moment in a molecule of water. (b) Shows the attraction between the negative charge of Cl⁻ and the partial positive charge of the dipole moment in a water molecule.

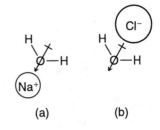

(a)   (b)

Before we continue, let's reiterate that a net molecular dipole, also called a **permanent dipole,** exists in a neutral molecule, where one end bears a partial positive charge (δ^+), and the other bears an equal and opposite partial negative charge (δ^-). In Chapter 3, we saw that the direction and magnitude of a dipole moment is represented by an arrow ($+\longrightarrow$), where the arrow's head represents the δ^- and the tail represents the δ^+. Therefore, a positive ion like Na⁺ will stick to the head of the arrow, and a negative ion like Cl⁻ will stick to the tail (Figure 7-1).

The concentrations of charge at the head and at the tail of a dipole moment are much smaller than that of an ion with a formal negative or positive charge; the charges within a dipole are only partial charges. Therefore, *ion-dipole interactions are weaker interactions than ion-ion interactions.*

The magnitude of a dipole moment is dependent upon the specific molecule. In general, the greater the magnitude of the dipole moment, the larger the concentration of charge at each end and the greater the strength of interaction between the dipole moment and an ion. For example, consider two different interactions—one between an Na⁺ ion and the dipole moment within a molecule of H_3C—F, and the second between that Na⁺ ion and the dipole moment within a molecule of H_3C—Cl. Because F has a greater electronegativity than Cl, we would expect that the dipole moment in H_3C—F is of greater magnitude than that in H_3C—Cl. Therefore, we expect the Na⁺ ion to stick better to a molecule of H_3C—F than to a molecule of H_3C—Cl. That is, we would expect the ion-dipole attractive interactions to be greater.

7.4 DIPOLE-DIPOLE INTERACTIONS

This intermolecular interaction exists between two neutral molecules. They can be two of the same molecule or different molecules, each with a permanent dipole. Those molecules can line up in two ways, such as to give an overall attractive interaction (Figure 7-2). In both orientations, note that the δ^+ of one dipole moment is near the δ^- of the other. Not surprisingly, both of those orientations are called **head-to-tail** orientations.

Figure 7-2 Two ways in which two molecules can be oriented so as to provide an overall attractive interaction.

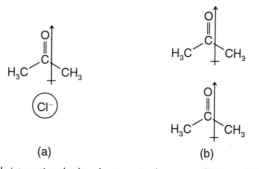

(a)   (b)

Figure 7-3 (a) Ion-dipole interactions lead to the attraction between a Cl⁻ ion and the dipole moment of a molecule of $(CH_3)_2C{=}O$. (b) Dipole-dipole interactions lead to the attraction between two molecules of $(CH_3)_2C{=}O$. The attractive forces in (a) are stronger than those in (b) because the $\delta+$ on $(CH_3)_2C{=}O$ is attracted much stronger to a full negative charge on Cl⁻ than a δ^- on another molecule of $(CH_3)_2C{=}O$.

As mentioned in Section 7.3, the concentration of charge at the head and tail of each dipole moment is smaller than that of an ion. Therefore, we can conclude that *the strength of attraction with dipole-dipole interactions is weaker than that with ion-dipole interactions.* For instance, a molecule of acetone, $(CH_3)_2C{=}O$, will stick better to a Cl⁻ ion than it will to another molecule of acetone (Figure 7-3).

In addition, dipole-dipole interactions become stronger if one or both of the dipole moments is increased in magnitude. So, because oxygen has a greater electronegativity than nitrogen, the dipole-dipole interactions are stronger between two molecules of H_2O than between two molecules of NH_3 (Figure 7-4).

7.5 HYDROGEN BONDING

Despite the fact that hydrogen bonding (abbreviated "H-bonding") is classified as a different intermolecular interaction, it is very closely related to the dipole-dipole interactions we examined in the previous section. In fact, many organic chemists treat hydrogen bonding

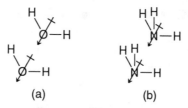

(a)   (b)

Figure 7-4 (a) Dipole-dipole attractive interactions between two molecules of H_2O. (b) Dipole-dipole attractive interactions between two molecules of NH_3. With the greater electronegativity of O, we would expect that the dipole-dipole interactions would be stronger in (a) than in (b).

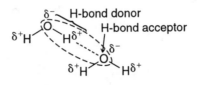

Figure 7-5 A hydrogen bond between two molecules of water. The hydrogen bond (dotted oval) is shown between the O—H of one molecule and the O of the other. The OH in the upper left molecule is the hydrogen-bond donor, and the O atom in the lower right molecule is the hydrogen-bond acceptor. The dashed line indicates an attractive interaction between the two molecules.

essentially as a subcategory of dipole-dipole interactions. However, there are two main differences: (1) Dipole-dipole interactions are defined as those between the *net molecular dipole moment* of one molecule and the *net molecular dipole moment* of another, whereas hydrogen bonding is between a *bond dipole* in one molecule and a *bond dipole* in another (see Chapter 3 to review the difference between these two types of dipoles); (2) specific atoms must be involved, and they must be arranged in a certain way to facilitate hydrogen bonding, or a **hydrogen bond.**

The specific arrangement required for a hydrogen bond is X—H---Y, where X and H are covalently bonded together, and Y is either an atom on a different part of the same molecule or is part of a separate molecule entirely. Notice, in particular, that the H atom is between X and Y; the presence of the hydrogen-bonding interaction (i.e., the hydrogen bond) is represented by the dotted line and implies an attraction between H and Y. A common mistake is to say that the X—H bond is the hydrogen bond in the previous scenario. Instead, *the hydrogen bond is the entire set of those three atoms in that arrangement.*

For that arrangement of atoms to form a hydrogen bond, X and Y must be either F, O, or N. The reason is due to the high electronegativity of those atoms. As a result, the X—H bond dipole is quite large in magnitude—that is, there is a large concentration of negative charge on the X atom and a large concentration of positive charge on the H atom. In addition, there should also be a large concentration of negative charge on the Y atom, given that Y also has a high electronegativity. Therefore, when the two molecules come together in a fashion yielding the arrangement X—H---Y, the large partial positive charge of the H is adjacent to the large partial negative charge of the Y, leading to a very stable situation. In other words, the molecules are stuck together rather strongly. Figure 7-5 demonstrates this point with a hydrogen bond between two molecules of H_2O, and Figure 7-6 demonstrates this with a hydrogen bond between a molecule of F—H and a molecule of $(CH_3)_2C{=}O$.

In every hydrogen bond that is formed, there is a hydrogen-bond donor and a hydrogen-bond acceptor. A **hydrogen-bond donor** is *the part of a molecule that contains the formal*

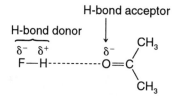

H-bond acceptor

H-bond donor

$\overline{\overset{\delta^-}{F}\,\overset{\delta^+}{H}}$------------$\overset{\delta^-}{O}$=C

CH$_3$

CH$_3$

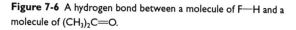

Figure 7-6 A hydrogen bond between a molecule of F—H and a molecule of $(CH_3)_2C$=O.

covalent bond between the highly electronegative X atom and the H atom. The **hydrogen-bond acceptor** is *just the electronegative atom Y.* In Figure 7-5, the donor would be the covalent O—H bond in the H_2O molecule in the upper left, and the hydrogen-bond acceptor would be the oxygen atom on the molecule in the lower right. In Figure 7-6, the hydrogen-bond donor would be the covalent F—H bond, and the acceptor would be the oxygen atom of the other molecule.

One reason we must discuss hydrogen-bond donors and acceptors is that not all molecules that contain F, O, or N can form hydrogen bonds. *If there are only hydrogen-bond acceptors, then no hydrogen bonds can be formed.* For example, no hydrogen bonds can form between two molecules of H_3C—F, between two molecules of $(CH_3)_2C$=O, or between one molecule of H_3C—F and one molecule of $(CH_3)_2C$=O. In all three of those scenarios, each molecule has a hydrogen-bond acceptor only. However, hydrogen bonds can be formed between a molecule of H_2C=O and a molecule of water (that is, H_2C=O---H—OH), given that there is a donor-acceptor pair.

Another reason to discuss hydrogen-bond donors and acceptors is to gain a sense of how strongly two hydrogen-bond forming molecules can stick together. We do so by counting the total number of hydrogen bonds that could potentially be formed between the two molecules, and then applying this general rule: *The greater the number of hydrogen bonds that can be formed between two molecules, the stronger the overall attraction between those molecules.* In terms of hydrogen-bond donors and acceptors, the greater the number of donor-acceptor pairs, the stronger the overall attraction between those molecules.

For example, if we consider two molecules of H_2O versus two molecules of ethylene glycol (HO—CH_2CH_2—OH, which is a product sold as antifreeze for your car), it would appear that more hydrogen bonding can exist between the two molecules of ethylene glycol. This is because there are more donor-acceptor pairs (Figure 7-7). Consequently, we would expect that two molecules of ethylene glycol would stick together much stronger than two molecules of water.

Before moving on, the last thing we must consider about hydrogen bonding is the relative strength of this type of interaction. First note that the hydrogen bond is really between a strong *partial* positive charge and a strong *partial* negative charge. Therefore, *in general, a hydrogen bond is weaker than ion-dipole interactions,* which are interactions

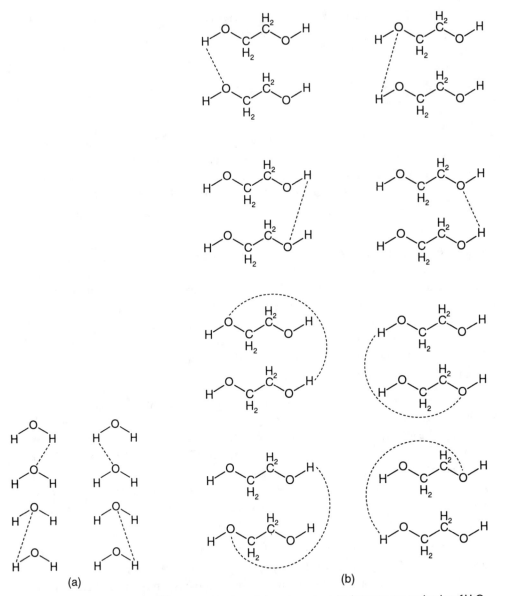

Figure 7-7 (a) Depicts all the different combinations of donor-acceptor pairs between two molecules of H_2O. (b) Depicts all the different combinations of donor-acceptor pairs between two molecules of ethylene glycol. Because there are several more combinations of donor-acceptor pairs in (b) than in (a), we expect that two molecules of ethylene glycol would stick together much stronger than two molecules of H_2O.

between a *partial* charge and a *full* charge. However, keep in mind that hydrogen bonding is closely related to dipole-dipole interactions but is special enough to bear its own name. Consequently, *hydrogen bonding is, in general, stronger than dipole-dipole interactions.*

7.6 DIPOLE–INDUCED DIPOLE INTERACTIONS

Dipole–induced dipole interactions exist between one molecule that has a net molecular (permanent) dipole, like HF, and another that does not, like H_2. But how could two molecules like these stick together? You might expect that there is no attraction that could exist, because a neutral molecule that has no permanent dipole has no region that bears excess charge—neither positive nor negative. Therefore, the partial positive charge on the molecule that has the permanent dipole doesn't see any excess negative charge on the other molecule. Likewise, the partial negative charge on the molecule with the permanent dipole doesn't see any excess positive charge on the other molecule. And this is, in fact, the case— at least initially. However, it turns out that the molecule with a permanent dipole *induces* a dipole (although a very weak and temporary one) on the molecule with no permanent dipole, and the subsequent interaction *becomes* a type of dipole-dipole interaction. The difference between this interaction and the dipole-dipole interactions we saw before is that *in dipole-dipole interactions, the interaction is between two permanent dipoles. On the other hand, in dipole–induced dipole interactions, the interaction is between a permanent dipole of one molecule and an induced (or temporary) dipole on the other.*

How does this induced dipole happen? The molecule with a permanent dipole approaches the other molecule in one of two ways—either with the positive end of the dipole pointed toward the nonpolar molecule, or with the negative end pointed toward the nonpolar molecule. If the positive end is pointed toward the nonpolar molecule, the electrons (which are negatively charged and are relatively free to move around) surrounding the nuclei of the neutral molecule will subsequently be attracted toward that positive end of the dipole. Therefore, the distribution of electrons in the nonpolar molecule has changed! There is a buildup of electrons (δ^-) on one side of the nonpolar molecule (the side closer to the partial positive charge of the molecule with the permanent dipole), and a deficiency of electrons (δ^+) on the other side (Figure 7-8). In other words, the net dipole from the polar molecule has *induced* a small, temporary dipole on the nonpolar molecule, and the nonpolar molecule has become **polarized.** The two molecules can then stick together, although

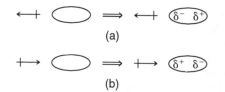

(a)

(b)

Figure 7-8 (a) On the left side of the double arrow, a molecule with a permanent dipole (like F—H) approaches a nonpolar molecule (oval, like H—H) with the positive end of the dipole pointed toward the nonpolar molecule. Shortly thereafter, on the right side of the double arrow, the nonpolar molecule has developed a small dipole, as a result of the polar molecule attracting electrons toward it and leaving a deficiency of electrons on the other side. (b) On the left side of the double arrow, the polar molecule approaches with its negative end. On the right side of the double arrow, a small dipole has formed as a result of the negative end repelling electrons to the other side of the nonpolar molecule.

weakly. A similar situation exists when the polar molecule approaches with its negative end, except that instead of attracting electrons to its side, it repels electrons to the other side.

Dipole–induced dipole interactions are weaker than the intermolecular interactions we previously looked at. This should make sense because dipole–induced dipole interactions are very similar to dipole-dipole interactions, but the dipole that is induced on the nonpolar molecule is very weak in magnitude. Therefore, we can immediately conclude that dipole–induced dipole interactions are weaker than dipole-dipole interactions.

7.7 INDUCED DIPOLE–INDUCED DIPOLE INTERACTIONS (A.K.A. VAN DER WAAL'S FORCES, OR DISPERSION FORCES)

To understand induced dipole–induced dipole interactions and their consequences, we must first better understand the behavior of electrons in a molecule that has no net dipole, such as H_2. In a molecule of H_2, there are two total electrons, which are shared between the two H nuclei. They are shared equally, because the H atoms are identical and therefore have identical electronegativities. On *average*, then, those electrons spend an equal amount of time around each of the two nuclei, which is what gives rise to the H_2 molecule having no permanent dipole.

However, it is *not* true that at every instant in time the two electrons are exactly opposite each other in the molecule. If we were to take a snapshot of the H_2 molecule at some instant in time, it may be that both electrons are very close to one nucleus and far away from the other nucleus (Figure 7-9). At that moment, there exists an **instantaneous dipole,** similar to the induced dipole we introduced in the last section. **Induced dipole–induced dipole interactions** are formed when that instantaneous dipole *induces* a dipole on another molecule, resulting, once again, in an interaction that resembles dipole-dipole interactions.

Figure 7-9 (Left) Two H_2 molecules (top and bottom) with the electrons evenly distributed. (Middle) At some instant in time, both electrons on one H_2 molecule may be very close to one nucleus, giving rise to an instantaneous dipole moment. (Right) The molecule with an instantaneous dipole has induced a dipole moment on the second H_2 molecule, resulting in a type of dipole-dipole attraction.

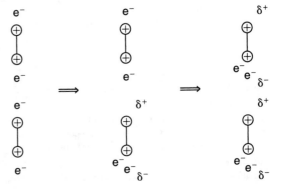

Although such interactions are typically quite weak, they are significant enough to worry about. As we will learn more about later, if it were not for these interactions, nonpolar molecules would only exist as gases and could never turn to liquids or solids, even at the coldest of temperatures—nothing would allow those molecules to stick together. On the contrary, we know that nonpolar molecules like H_2, O_2, and N_2 *do* liquefy if the temperature is cold enough. For example, N_2 liquefies at around $-200°C$ (the cold temperatures that are required are a testament to how weak this type of interaction really is).

Keep in mind two things about induced dipole–induced dipole interactions. First, *induced dipole–induced dipole interactions exist between any two atomic or molecular species that contain electrons*. This is because of what gives rise to induced dipole–induced dipole interactions—the distortion of those electrons relative to positively charged nuclei.

Second, induced dipole–induced dipole interactions are usually very weak—in general weaker than dipole–induced dipole interactions. This is because in dipole–induced dipole interactions, the dipole that does the inducing is a permanent dipole, which can be rather large in magnitude. In induced dipole–induced dipole interactions, the dipole that does the inducing of the second dipole is itself an induced dipole, which is often much smaller in magnitude than a permanent dipole.

We must be careful when comparing the strength of induced dipole–induced dipole interactions to other intermolecular interactions. This is because *the strength of induced dipole–induced dipole interactions in general increases with the number of electrons in each species*. Therefore, whereas the induced dipole–induced dipole interactions are quite weak for the interaction between two molecules of CH_4 (only 10 total electrons), they are much stronger for the interaction between two molecules of $CH_3(CH_2)_{16}CH_3$ (146 total electrons)— even stronger than the hydrogen bonding that exists between two molecules of water!

7.8 SUMMARY OF INTERMOLECULAR INTERACTIONS

Each of the intermolecular interactions we have learned about in this chapter was introduced independent of the other intermolecular interactions. Rarely, however, can we say that one and only one intermolecular interaction is in effect. Under most circumstances where there are attractive forces between two molecules, there is more than one intermolecular interaction present. In fact, typical organic chemistry problems often force you to consider multiple intermolecular interactions simultaneously. Some ask you to determine exactly which of the six intermolecular interactions exist between a pair of given molecules. Others have you determine the relative strengths of the intermolecular interactions that are active.

As an example, suppose we are asked to determine what intermolecular interactions exist between a molecule of NH_3 and a molecule of H_2O. First, we see that hydrogen bonding can exist between them. Either the O—H of the H_2O molecule can be the H-bond donor with the N of the NH_3 as the H-bond acceptor, or the N—H can be the donor with the O atom the acceptor. Second, both molecules have permanent dipoles; therefore, dipole-dipole interactions should also be present. Third, realize that the permanent dipole of the H_2O molecule can push around the electrons of the NH_3 molecule. As a result, the dipole of H_2O will induce a dipole on NH_3, giving rise to dipole-induced dipole interactions.[1] Similarly, the permanent dipole of the NH_3 molecule can induce a temporary dipole moment on the H_2O molecule. Fourth, both neutral molecules have electrons—therefore, induced dipole–induced dipole interactions should also be active. In all, then, there are four types of intermolecular interactions that are simultaneously active to help a molecule of H_2O and a molecule of NH_3 stick together.

Another type of question may have you go one step further, asking you to determine how strongly two molecules stick together, in a relative sense. For example, you might be asked which pair of molecules sticks together better, either H_2O with NH_3, or H_2O with CH_4. To answer this question, just do two things. First, for each pair of molecules, determine exactly which intermolecular interactions are active. Second, figure out which of those interactions is going to be the most important—that is, the strongest. We can assume that the strongest active intermolecular interaction will overshadow the weaker ones that are present. In other words, *the majority of how well a pair of molecules sticks together is from the strongest intermolecular interaction that is present.*

We have already identified the four intermolecular interactions active in the first pair of molecules; of those, the hydrogen-bonding interaction is the most important to consider. In the second pair of molecules, there can be no hydrogen bonding. Furthermore, because CH_4 is a nonpolar molecule, there can be no dipole-dipole interactions—those interactions that do exist are only dipole–induced dipole interactions and induced dipole–induced dipole interactions. The most important of the interactions between CH_4 and H_2O are therefore dipole–induced dipole interactions, which are much weaker than hydrogen bonding. Therefore, we can say that a molecule of H_2O will stick to a molecule of NH_3 better than it will to a molecule of CH_4.

7.9 APPLICATION: MELTING POINT AND BOILING POINT DETERMINATION

You will invariably face homework and exam problems that give you the Lewis structures of several molecules and ask you to rank those molecules in order of boiling point or melting point (the respective temperatures at which boiling or melting occurs). One way to

---

[1]Note that the isolated NH_3 molecule already has a permanent dipole moment; therefore, the H_2O molecule will serve to alter that dipole slightly when the two molecules are close together.

attack such a problem is to apply the various tricks that you memorized from somewhere else. Wrong! The easier and better way is to apply what you have learned in the previous sections of this chapter, as well as what you know about the processes of melting and boiling.

Let's first write out what we know about melting and boiling. Both processes are nothing more than *changes of phase*—melting is from solid to liquid, and boiling is from liquid to gas (or vapor). They are *not* chemical reactions. Covalent bonds are *not* broken and/or formed. So, for example, when ice melts, we say that solid H_2O turns into liquid H_2O, but ALL of the H_2O molecules remain intact during that process. We also know from experience that both of those processes require heat.

In addition, we know something about what it means to be a solid, a liquid, and a gas. In a solid, molecules are in very close contact, essentially touching one another. They are also in a very nicely ordered array—what we call either a lattice or a crystal—and are *frozen in place*. The molecules do not **translate** (i.e., move from one place to another), nor do they rotate in place.

Liquids, like solids, have molecules in very close contact and are essentially touching one another. Unlike solids, however, molecules in a liquid can rotate in place and translate.

Gases are essentially isolated molecules—they basically don't see or feel any other molecules. Like liquids, gaseous molecules can also translate and rotate.

Understanding this, and what we learned about intermolecular interactions, we can explain why it is that the process of melting requires heat energy. Solids are comprised of molecules that are frozen in place in a highly ordered arrangement, which is the best scenario to maximize the attractive intermolecular interactions between molecules. Figure 7-10 demonstrates this with dipole-dipole interactions specifically, where the molecules are very nicely ordered to give the most head-to-tail interactions that could possibly exist. If that solid substance were melted, the translation and the rotation of the molecules in the liquid phase must result in the destruction of at least some of those good head-to-tail interactions

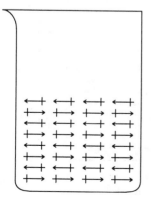

Figure 7-10 A collection of several of the same molecule that has a net molecular dipole. The high order of the system depicts this substance as a solid

Figure 7-11 Similar to Figure 7-10, representing the substance as a liquid, with more disorder.

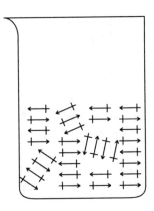

(Figure 7-11). That costs energy! And in order to supply that energy, heat must be provided.

If we go one step further and boil the substance, converting it from liquid to gas, then all of the molecules become isolated, and ALL of the remaining good head-to-tail interactions are destroyed (Figure 7-12). That costs even more energy in the form of heat, and that is why boiling points are higher than melting points!

Now that we know that melting involves the destruction of some percentage of the good interactions between molecules, and boiling involves the destruction of all of the rest, we can rank melting points and boiling points, given only the Lewis structures of a set of molecules. It works this way: If there is some strength of attraction between a pair of molecules, then melting will require a certain amount of heat. If that strength of attraction is increased, then melting will require more heat, and the melting point is higher—the substance requires higher temperatures to melt. Similarly, if there is some strength of interaction between molecules that gives rise to a certain boiling point, then increasing that

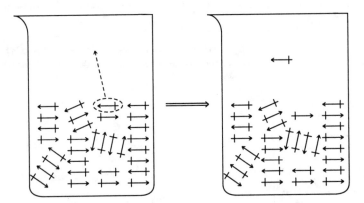

Figure 7-12 The process of boiling. One molecule from the liquid is shown to have been removed and converted to the gas phase, where all possible good interactions have been destroyed.

strength of interaction will increase the boiling point. The last piece of the puzzle is simply applying what you know about the relative strength of attraction between a pair of molecules resulting from the previous intermolecular interactions.

Let's look at some examples. Suppose we were asked to determine which molecule has the higher boiling point, CH_4 or CH_3F. We must determine how much heat energy it takes to convert liquid CH_4 to gaseous CH_4, and compare that to the heat energy required to convert liquid CH_3F to gaseous CH_3F. For the former, we realize that the interactions that are broken in going from liquid to gas are those between two molecules of CH_4, which are just induced dipole–induced dipole interactions (CH_4 is nonpolar). For the latter, the interactions that are broken are between two molecules of CH_3F, the most important of which are dipole-dipole interactions. Because dipole-dipole interactions are much stronger, we would expect CH_3F to have a significantly higher boiling point than CH_4.

Suppose, now, that we were asked to determine the relative melting points of CH_4 and C_2H_6. The intermolecular interactions that exist between two molecules of CH_4 are only induced dipole–induced dipole interactions. The same is true for two molecules of C_2H_6. The difference is that C_2H_6 has more electrons. As a result, induced dipole–induced dipole interactions are stronger between two molecules of C_2H_6, which therefore gives rise to a higher melting point for C_2H_6 than CH_4. This idea can be generalized to say that, *all else being equal, more massive molecules (because they have a greater number of electrons) have higher boiling and melting points.*

7.10 APPLICATION: SOLUBILITY

As with boiling points and melting points, let's first make sure that we understand the process of dissolving before we try to say anything about solubility. When one pure substance (the **solute**) dissolves into another pure substance (the **solvent**), the first simply mixes with the second to yield a solution. A **solution** is a homogeneous mixture of the solute and the solvent (although the solvent can technically be any of the three phases—solid, liquid, or gas—organic chemistry deals exclusively with the solvent being a liquid). One thing we can say immediately about the process of dissolving is that it will occur spontaneously only if the two substances want to mix. That sounds trivial, but we must look more closely at what it means for two substances to *want* to mix, which entails a closer look at what is happening with the intermolecular interactions during the process.

Before combining two substances together, the intermolecular interactions that exist in each of the pure substances are between two identical molecules. If one substance is liquid H_2O, the most important intermolecular interaction is from hydrogen bonding. If the second substance is liquid $CH_3(CH_2)_4CH_3$, the predominant intermolecular interaction is induced dipole–induced dipole interactions. If the two substances were then combined

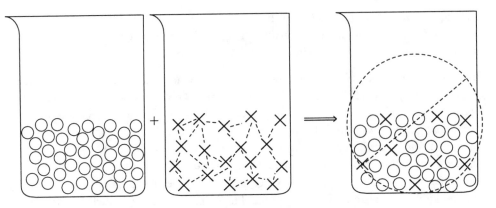

Figure 7-13 (Left) A nonpolar substance is combined with a substance that can undergo hydrogen bonding. The result (right) is the destruction of the hydrogen bonding interactions that exist when the substances are pure and unmixed. Consequently, the two substances will unmix and form two layers.

together (Figure 7-13), then molecules of $CH_3(CH_2)_4CH_3$ would become interspersed with molecules of H_2O, which means that we must consider the intermolecular interactions that exist between the two different types of molecules. Because H_2O is polar and $CH_3(CH_2)_4CH_3$ is nonpolar, the most important intermolecular interaction between the molecules is dipole–induced dipole interactions, which we know is significantly weaker than H-bonding. So, it appears that in the mixture, the nonpolar molecules of $CH_3(CH_2)_4CH_3$ serve to destroy the much better H-bonding interactions that existed when the two substances were separated. Therefore, if these two substances were combined and vigorously shaken to force them to mix, there would be a strong driving force to *unmix*. And that is exactly what would happen! The two substances would separate into two different layers—one on top of the other, with the more dense one being on the bottom; and we would say that $CH_3(CH_2)_4CH_3$ is not soluble in H_2O, or that the two compounds were not **miscible**. In fact, this is why Italian dressing tends to form two layers when it sits—an oil layer forms on top of a water (vinegar) layer.

Let's now look at what would happen if we tried to mix CH_2Cl_2 with $(CH_3)_2C{=}O$. Both molecules are polar. In pure CH_2Cl_2, the most important intermolecular interactions are dipole-dipole interactions. The same is true for pure $(CH_3)_2C{=}O$. If the two substances were mixed (Figure 7-14), molecules of CH_2Cl_2 would become interspersed with molecules of $(CH_3)_2C{=}O$, and the intermolecular interactions between those two molecules would also be dipole-dipole interactions. Therefore, to first approximation, the situation in which the two substances are mixed appears to be no better and no worse than when the two substances were pure and separated. So, will these two substances remain mixed or will they unmix?

They will remain mixed—because the most important factor in determining solubility (i.e., intermolecular interactions) is essentially a wash in this case. What we have not yet considered, however, is something called **entropy**, which is often described as *a measure of*

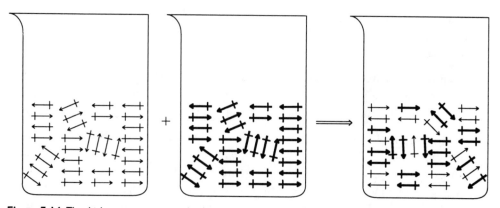

Figure 7-14 The thick arrows represent dipole moments from one kind of molecule, and the thin arrows represent dipole moments from a second. Good head-to-tail interaction exists in both the pure substances (left) and in the mixed solution (right). The two substance will remain mixed because of increased entropy (disorder) in the solution.

disorder in a system. One of the fundamental laws of chemistry and physics is that, all else being equal, *systems tend to want to be more disordered*, or, have more entropy (just think of your bedroom ...). In the case we are considering, all else is essentially equal, and the more disordered system is that in which the two substances are mixed together. Therefore, we would say that CH_2Cl_2 and $(CH_3)_2C{=}O$ would mix, and one is soluble in the other. That is, the two substances would be miscible.

In another scenario, we could mix together two nonpolar substances (substances whose molecules possess no net molecular dipole) such as hexane, $CH_3(CH_2)_4CH_3$, and carbon disulfide, $S{=}C{=}S$. In this case, there will be very weak induced dipole–induced dipole interactions in each of the pure substances, and, likewise, the same type of weak interactions in the mixture (Figure 7-15). Therefore, because we don't destroy or gain any

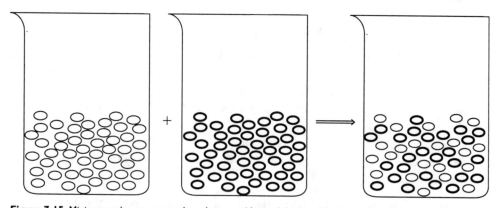

Figure 7-15 Mixing together two nonpolar substances (thin and thick ovals) does not destroy any strong intermolecular interactions. Increased entropy in the solution allows the two substances to remain mixed.

strong interactions in the process of mixing, the driving force to increase entropy becomes the most important factor. The end result is that two nonpolar substances will be miscible.

From what we have said so far, one can easily see the origin of the phrase "like dissolves like." In general, two polar compounds will be miscible in each other, as will two nonpolar compounds. But a polar compound and a nonpolar compound are generally immiscible.

Applying the rule of thumb that "like dissolves like," we can then say that hydrogen-bonding compounds will be soluble in other compounds that can undergo hydrogen bonding. For example, H_2O and $HOCH_2CH_2OH$ are miscible. The reason here is that significant hydrogen bonding can exist between the two different molecules, just as in the pure separated substances. This is another situation in which the interactions in the mixture are about as good as when separated, so that increased entropy becomes the important factor.

What about CH_2Cl_2 and CH_3CH_2OH? Using "like dissolves like," we might be tempted to say that the two are not miscible, because the first molecule cannot participate in hydrogen bonding, whereas the second one can. However, it turns out that the two are significantly soluble in each other. The reason is that both have fairly strong permanent dipole moments. Therefore, when mixed together, the two molecules would attract each other through dipole-dipole interactions. Yes, those dipole-dipole interactions that would be in effect when the two substances are mixed are weaker than the hydrogen bonding that would exist in the pure sample of CH_3CH_2OH. This suggests that the two substances should unmix. However, dipole-dipole interactions are fairly strong. Moreover, staying mixed has entropy working in its favor.

Understanding all of this, we can then ask, "What kinds of solvents are ionic compounds typically soluble in?" For ionic compounds to be soluble, it must be that ionic bonds (from the pure substance) are broken, which, as we previously argued, must require a LARGE amount of energy. Therefore, whatever interactions that exist in the mixed solution must be sufficient to overcome the loss of those ionic bonds. Ion-dipole interactions are sufficient; the dipole can be either a net molecular dipole, or it can be a strong bond dipole from a hydrogen-bond donor such as O—H.

The reason ion-dipole interactions can overcome ion-ion interactions is not because the strength of an individual ion-dipole interaction is greater than an ion-ion interaction (ionic bond)—in fact, we know that the reverse is true. Instead, it is because for each ionic bond that must be broken, *several* ion-dipole interactions can exist (Figure 7-16). Therefore, in general, polar solvents (like water and CH_2Cl_2) can dissolve ionic compounds. Such a phenomenon that exists in the mixed solution (the surrounding of ions by the dipoles of solvent molecules) is called **solvation** and is important in **solvent effects**—the topic of the next section of this chapter.

When ionic compounds do dissolve, they dissolve as their constituent ions. Most of us are familiar with NaCl dissolving as $Na^+ + Cl^-$, and NaOH dissolving as Na^+ and OH^-.

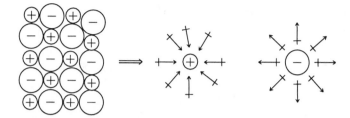

Figure 7-16 Dissolving an ionic compound in a polar solvent. The ions in an ionic compound are held together by very strong ionic bonds. Each ion-dipole interaction is weaker than an ionic bond, but several of the molecules with dipoles can surround each ion, providing a better overall interaction. This is called *solvation* of the ions.

However, one common difficulty that students have is in dealing with compounds such as $KOCH_3$ or $NaNH_2$ dissolved in solution. But the last two are no different from the first two—they are ionic compounds that dissolve as $K^+ + CH_3O^-$, and $Na^+ + H_2N^-$, respectively. Furthermore, the reactive species are the anions, which can act as both bases and nucleophiles (remember, metal cations like Li^+, Na^+, and K^+ are unreactive and tend to act as spectator ions).

7.11 APPLICATION: SOLVENT EFFECTS ON NUCLEOPHILES

In an application at the end of Chapter 5, you learned how to determine the relative strength of a nucleophile, based on only its Lewis structure. In that application, we applied what we knew about relative stabilities of charges to determine relative strengths of nucleophiles. However, in determining the relative nucleophile strengths in this manner, we did not consider the role solvent molecules might play. And these roles of solvents—**solvent effects**—are important because most organic reactions take place in some variety of solvent.

There are essentially two types of solvents that we must consider when regarding nucleophile strengths: (1) protic solvents and (2) aprotic solvents. **Protic solvents** are defined as *solvents that have an H-bond donor*, which include H_2O, CH_3OH, CH_3CH_2OH, $CH_3CH_2NH_2$, etc. In each of those molecules, the H that is covalently bonded to the O or N bears a large δ^+, as a result of the high electronegativity of O and N. Therefore, if a negatively charged nucleophile, like Cl^-, is dissolved in a protic solvent, like H_2O, then the H_2O molecules will stick to the Cl^- ion VERY strongly as a result of the interaction between the full negative charge on Cl^- and the large δ^+ on the H in H_2O. The end result is that one Cl^- ion will have several H_2O molecules strongly attached to it (Figure 7-17). In other words, the Cl^- is *strongly solvated* by H_2O.

Figure 7-17 A Cl⁻ ion is very strongly solvated by molecules of H₂O, as a result of the large δ^+ on each H covalently bonded to an O atom.

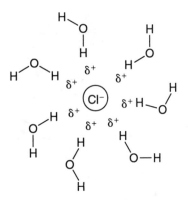

Recall from Chapter 5 that the strength of a nucleophile stems from its ability to seek out a center of positive charge. In order for a nucleophile to seek out anything, it must be able to move. Think about the mobility of a Cl⁻ ion that is dissolved in a protic solvent like H₂O (Figure 7-17). The water molecules that are strongly stuck to the Cl⁻ ion substantially hinder its mobility and therefore dramatically decrease its strength as a nucleophile. This concept is particularly important to understand for nucleophilic substitution reactions that we will examine in Chapter 8.

You can imagine that this solvent effect of *weakening the nucleophile is actually worse for smaller sized anions*, where the concentration of negative charge is typically greater. With a greater concentration of negative charge, the δ^+ from the H on the protic solvent will stick even better. In fact, *what may occur as a result of this is a change in relative nucleophile strength.*

For example, free of solvent, we would normally say that Cl⁻ is a better nucleophile than Br⁻, because, due to its larger size, Br⁻ can better accommodate the negative charge and is therefore more stable. However, if you dissolve both of these ions in water, the water molecules will stick to the Cl⁻ much stronger than they will to Br⁻, due to the higher concentration of negative charge on the much smaller Cl⁻. Therefore, Cl⁻ is weakened much more than Br⁻. The end result is that in water, the relative strengths of these nucleophiles are actually reversed! Both remain strong nucleophiles because of their formal −1 charge, but Br⁻ is a stronger nucleophile than Cl⁻. This is a general trend within a column of the periodic table. Therefore, as another example, HS⁻ is a stronger nucleophile than HO⁻, despite the fact that free of solvent, HO⁻ is intrinsically stronger as a nucleophile.

Reversals of nucleophile strength tend not to occur for atoms in the same row of the periodic table bearing the same formal charge. For example, H₂N⁻ is a stronger nucleophile than HO⁻ both in protic solvents and free of solvent altogether. The reason is that atoms in the same row tend to be similar in size. Therefore, if two atoms in the same row bear the same formal charge, the concentration of charge on those atoms is roughly the same. Consequently, they will be solvated by roughly the same amount, leaving their stability relative to each other unchanged.

Reversals of nucleophile strength also tend not to occur for atoms bearing different formal charge. As an example, we know that free of solvent, H_2N^- is a stronger nucleophile than H_3N because the formal negative charge in H_2N^- is inherently unstable. In a protic solvent like water, the weakening of H_2N^- as a nucleophile is much more pronounced than the weakening of H_3N, because the concentration of negative charge is much stronger on the N atom of H_2N^-. However, the additional solvation of H_2N^- is not enough to make it a weaker nucleophile than H_3N—the instability of the formal negative charge is too great for this to happen.

The story is a little different in **aprotic solvents**—*solvents that do not form hydrogen bonds.* Common aprotic solvents you will encounter include acetone [$(CH_3)_2C=O$], dimethyl sulfoxide [DMSO, $(CH_3)_2S=O$], and dimethyl formamide [DMF, $(CH_3)_2N-CH=O$]— see Figure 7-18. Notice that in each of these compounds, there is no H atom covalently bonded to one of the three highly electronegative atoms (N, O, F). They do, however, all have net molecular dipoles. Therefore, as a result of ion-dipole interactions, each of these solvent molecules will stick fairly well to an anion like Cl^-—hence ionic compounds are capable of dissolving in these solvents. But those solvent molecules will not stick nearly as well as protic solvents would. Therefore, aprotic solvents tend not to weaken the nucleophile as greatly as protic solvents do. That is, solvation by aprotic solvents is much weaker than solvation by protic solvents. *We therefore tend not to see reversals of nucleophilicity in aprotic solvents.*

As a final note to this chapter, it is very important that you understand and keep in mind the different effects caused by the different types of solvents and WHY they are different. Failure to understand this is often a source of major confusion when it comes to understanding nucleophilic substitution reactions and elimination reactions—which we will cover extensively in the next chapter.

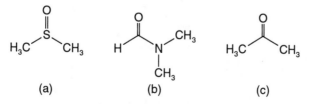

Figure 7-18 Common aprotic solvents. (a) Dimethyl sulfoxide (DMSO). (b) Dimethyl formamide (DMF). (c) Acetone.

Problems

7.1 Rank the following in order of increasing solubility in acetone $[(CH_3)_2C=O)]$. (Hint: Look for pairs of similar molecules, and for each pair, determine which one is more soluble.)

a $CH_3CF_2CH_3$ **b** CO_2 **c** $HOCH_2CH_2CH_2OH$
d O_2 **e** $CH_3CHClCH_3$ **f** CS_2
g CH_3CHFCH_3 **h** $CH_3CH_2CH_2OH$

7.2 For each pair of species, which is a stronger nucleophile in DMSO?

a CH_3OH vs. CH_3O^- **b** CH_3CH_2OH vs. CH_3CH_2SH
c H_2S vs. HS^- **d** CH_3Li vs. $NaOH$
e Cl^- vs. I^- **f** HS^- vs. Cl^-
g H_2P^- vs. I^- **h** $CH_3CH_2NH_3^+$ vs. $CH_3CH_2NH_2$

7.3 Repeat Problem 7.2 using H_2O as the solvent.

7.4 Rank the following in order of increasing boiling point.

a $CH_3CH_2CH_2OH$ **b** CH_3CO_2H **c** $(CH_3)_2C=O$
d $CH_3CH_2CH_2CH_3$ **e** CH_4 **f** $HO_2CCH_2CO_2H$
g $CH_3CH_2CH_2NH_2$

7.5 In general, an "enol" is unstable in solution. An enol is a functional group in which an alk*ene* functional group directly bonded to a carbon atom of an alc*ohol*—that is, $C=C-OH$. Most enols spontaneously undergo a rearrangement to $HC-C=O$. The enol shown here, however, is more stable than usual. Explain.

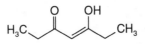

7.6 Which of the following nucleophiles will be weakened more in a protic solvent? Why?

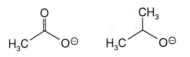

$S_N1/S_N2/E1/E2$ Reactions: The Whole Story

8

8.1 INTRODUCTION

There are two major reasons for this chapter. First, the S_N1, S_N2, E1, and E2 reactions are closely related to one another, and, for most students, they turn out to be the most difficult set of reactions in organic chemistry. They are typically encountered in the middle of the first semester and are integral in whether or not a student does well in the course. Second, those four reactions bring together all of the major concepts we have learned thus far. As you read through this chapter, keep note of how we incorporate aspects of bonding, isomerism (in particular, stereoisomerism), mechanisms, and intermolecular interactions (especially solvent effects).

These reactions tend to be challenging because, in general, if one reaction occurs, then *all four will occur simultaneously* (although at different reaction rates). Why is this so? The answer lies in the similarities among all four mechanisms (Figure 8-1). Each reaction involves a substrate containing a leaving group. In addition, even though two of the mechanisms involve a nucleophile and the other two involve a base, species that are *nucleophiles can also act as bases, and vice versa.*

The reason for the dual nature of bases and nucleophiles is as follows. A base contains an atom with excess electron density (partial or full negative charge) and a lone pair, allowing it to form a bond with an electron-poor proton (H^+). A nucleophile contains an atom with excess electron density and a lone pair, allowing it to form a bond with an electron-poor non-H atom (usually a C atom), often displacing a leaving group in the process. Therefore, whether we call a species a base or a nucleophile depends entirely on the reaction it undergoes.

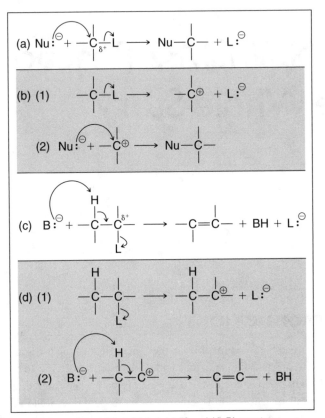

Figure 8-1 Complete mechanisms of the (a) S_N2, (b) S_N1, (c) E2, and (d) E1 reactions.

For example, HO^- acts as a base in Equation 8-1a, and acts as a nucleophile in Equation 8-1b. Notice that in both cases, whether the species acts as a base or as a nucleophile, a curved arrow originates from that electron-rich species and points to the species that is electron poor (the δ^+ on either the H of the acid or C of the substrate R—L).

$$H{-}\overset{..}{\underset{..}{O}}{:}^{\ominus} \quad H{-}A \longrightarrow H_2O + :A^{\ominus} \tag{8-1a}$$

$$H{-}\overset{..}{\underset{..}{O}}{:}^{\ominus} \quad {-}\overset{|}{\underset{|}{C}}{-}L \longrightarrow HO{-}\overset{|}{\underset{|}{C}}{-} + :L^{\ominus} \tag{8-1b}$$

Despite the fact that all four substitution and elimination reactions occur simultaneously, there is often a **major product,** which is the one that is *formed in the greatest amount.* You will be expected to determine the major product of these reactions. To do so, you must first determine the *predominant mechanism* (i.e., the mechanism that proceeds the fastest) that occurs under the specific set of reaction conditions that you are given. Keep in mind, though, that *the predominant mechanism is different under different conditions!* Once you have determined the predominant mechanism, predicting the products is only a matter of working through that mechanism, as you did in Chapter 6.

Determining the predominant mechanism is usually the single most difficult aspect of the S_N1, S_N2, E1, and E2 reactions. This is because there are *five factors* that dictate how favorable each reaction is. For a given set of reaction conditions, it is your job to figure out which reaction each of the five factors favors and then weigh everything to determine the predominant reaction.

By and large, students' frustration with these reactions stems from trying to memorize which set of conditions favors which reaction. Textbooks usually add to the temptation, providing a table that summarizes such things. *Don't attempt to memorize it!* Doing so without understanding WHY leads to disaster. Instead, throughout the remainder of this chapter, you will learn that the mechanisms in Figure 8-1 lead to a rate equation for each reaction. The rate equations, along with a fundamental understanding of what drives each mechanism, provide insight into how to use the five factors for each reaction. Toward the end of this chapter, we will work through a number of different scenarios in order to gain experience in applying the five factors.

8.2 RATE EQUATIONS

The mechanisms for the S_N2, S_N1, E2, and E1 reactions are presented in Figure 8-1. In the S_N2 mechanism, a nucleophile displaces a leaving group on the substrate in a single step. Therefore, the nucleophile can be thought of as *forcing the action—it forces the leaving group to leave* (it is important that you can see this for what we will discuss shortly).

The S_N1 mechanism, on the other hand, occurs in two steps. First the leaving group leaves, leaving behind a carbocation (R^+). The electron-rich nucleophile then adds to the electron-poor carbocation. *The first step of the S_N1 mechanism must be significantly slower than the second step*, because in the first step, a stable molecule with no charge reacts to form two charged species, which are much less stable ("charge is bad!"). This will occur on its own, but very slowly. In the second step, two unstable charged species react very quickly to form a stable molecule with no charge. Think about the S_N1 mechanism as the nucleophile *waiting for the leaving group to leave* before it attacks (again, it is important that you can see this).

Similarly, the E2 mechanism occurs in one step, whereas the E1 mechanism occurs in two steps. In the E2 mechanism, a base pulls off an H^+ and the leaving group leaves simultaneously. As in the S_N2 reaction, *the base in the E2 mechanism can be thought to force the action*—it pulls off the proton, which forces the leaving group to leave.

The first step of the E1 mechanism is parallel to that in the S_N1 mechanism, where the leaving group leaves spontaneously. In the second step, a base pulls off the proton, and the double bond is formed to complete the products. We can determine the relative rates of reaction for the first and second steps of the E1 mechanism, just as we did for the S_N1 mechanism. In the first step, a stable neutral molecule breaks apart to form two unstable charged species—this step will therefore be slow (in comparison to the second step). In the second step, two unstable and reactive charged species produce a stable neutral species— this step will therefore be fast (in comparison to the first step). As with the S_N1 mechanism, think about the E1 reaction as the base *waiting for the leaving group to leave* before it pulls off the H^+.

As we mentioned in the introduction, it is important to be able to say something about the rate of the overall reaction for each mechanism. This can be done straightforwardly with the mechanisms (Figure 8-1). In the S_N2 mechanism, the entire reaction occurs in one step. In that step, there are two reactant molecules that come together to form products. The rate at which this step occurs, then, depends on how much of each type of molecule is present. In other words, *the rate of the reaction is dependent on the concentration of the nucleophile, as well as the concentration of the substrate,* according to Equation 8-2 (concentration is indicated by brackets—[ ]). Specifically, the rate of the reaction is proportional to the product of the concentration of the nucleophile and the concentration of the substrate—k_{S_N2} is a constant, called the **rate constant,** for the reaction. Therefore, the rate of the reaction increases by increasing either the concentration of the nucleophile or the concentration of the substrate. This should make sense, because the greater the concentration of the nucleophile, the greater the number of nucleophile molecules present in solution, and the more often a nucleophile will come together with a substrate to form product. By the same token, the greater the concentration of the substrate, the greater the number of substrate molecules present, and the more often a substrate molecule will come together with a nucleophile to form product.

$$\text{Rate}(S_N2) = k_{S_N2}[\text{Nu}][\text{R—L}] \qquad (8\text{-}2)$$

A very similar story can be told for the E2 mechanism, because of the similarity of its mechanism to that of the S_N2 mechanism. Namely, the E2 reaction occurs in one elementary step. *The rate of the E2 reaction therefore depends on the concentration of both reactants,* according to Equation 8-3—k_{E2} is the rate constant. The greater the concentration of the base, the more often a molecule of the base will come together with a molecule of the substrate to form product. Similarly, the greater the concentration of the substrate, the more often a molecule of the substrate will come together with a molecule of the base to form product.

$$\text{Rate}(E2) = k_{E2}[\text{Base}][\text{R—L}] \qquad (8\text{-}3)$$

The S_N1 and E1 mechanisms are somewhat different, because each is a two-step mechanism. The equations that describe the rates of their overall reactions will therefore not take exactly the same form as Equations 8-2 and 8-3. To derive those rate equations, we must first discuss the concept of a rate determining step of a mechanism.

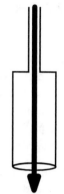

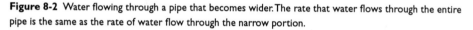

Figure 8-2 Water flowing through a pipe that becomes wider. The rate that water flows through the entire pipe is the same as the rate of water flow through the narrow portion.

A **rate determining step** is *the elementary step of a mechanism that dictates the rate of the overall reaction;* the rate of the overall reaction is essentially the same as that of the rate determining step. Not all mechanisms have a rate determining step, but both the S_N1 and the E1 mechanisms do. For these mechanisms, *the rate determining step is the slowest step—* the so-called "bottleneck" of the reaction. Think about this as water traveling through a pipe, where the volume that travels through the pipe per second is analogous to the rate of the overall reaction.

Imagine, then, a pipe that is initially quite skinny, but at some point widens significantly (Figure 8-2). The skinny portion of the pipe represents the slow step of a two-step mechanism, and the wider portion represents the fast step. Even though the wider portion of the pipe could sustain larger volumes of water per second, the rate at which water ends up traveling through the entire pipe (analogous to the overall reaction) is dictated by how much water per second comes through the thinner portion (the slow step of the reaction mechanism). Specifically, the rate at which water travels through our set of two pipes is the same as the rate at which water travels through the small pipe. Similarly, the rate of an overall reaction is the same as the rate of the rate determining step (i.e., the slow step).

Realizing this, to determine the rate of the overall reaction, we must only determine what dictates the rate of the slow step. For the S_N1 mechanism (Figure 8-1), we observe that there is only one reactant molecule in the first step (i.e., the slow step, or rate determining step). That one molecule is the substrate. Therefore, similar to our treatment of the previous two mechanisms, we can say that *the rate of the S_N1 reaction is dictated by the concentration of the substrate only, and NOT the nucleophile (Equation 8-4). For the same reasons, the rate of the E1 reaction is dictated by the concentration of the substrate only (Equation 8-5).* The difference in the two equations is in the rate constants.

$$\text{Rate}(S_N1) = k_{S_N1}[\text{R—L}] \qquad (8\text{-}4)$$

$$\text{Rate}(E1) = k_{E1}[\text{R—L}] \qquad (8\text{-}5)$$

8.3 FACTOR #1: STRENGTH OF NUCLEOPHILE/BASE

As discussed previously, one of the key differences between the S$_N$1 and S$_N$2 reactions is what drives them. The S$_N$2 reaction is primarily driven by the nucleophile displacing the leaving group—that is, the nucleophile forces the action. Therefore, *the stronger the nucleophile, the faster the rate of the S$_N$2 mechanism.* Another way to see this is that the nucleophile shows up in the rate equation (Equation 8-2), suggesting that the nucleophile's identity is important—the stronger it is, the faster the reaction.

In the S$_N$1 mechanism, on the other hand, the nucleophile essentially sits back and waits for the leaving group to come off before it adds in to the highly reactive carbocation. As a result, *changing the strength of the nucleophile has essentially no effect on the rate of the S$_N$1 reaction.* Whether it is a strong or a weak nucleophile, the S$_N$1 mechanism has the nucleophile sit back and wait for the leaving group to come off before the nucleophile adds in. A second way of justifying this is that the nucleophile does not appear in the rate equation (Equation 8-3). Therefore, the nucleophile's identity is not important.

The lesson to learn from this is that strong nucleophiles tend to favor S$_N$2 reactions, and weak nucleophiles tend to favor S$_N$1 reactions. Certainly, if a nucleophile is strong, then it will help speed up the S$_N$2 reaction considerably, but it will not speed up the S$_N$1. On the other hand, if the nucleophile is weak, then it cannot force off the leaving group, which means that the S$_N$2 mechanism cannot happen. However, a weak nucleophile can still sit back and wait for the leaving group to come off by itself and undergo the S$_N$1 mechanism.

Given their different tendencies, it is important to be able to recognize strong nucleophiles and weak nucleophiles. *Although there is no clear cutoff between them, it is relatively safe to say that a nucleophile bearing a full negative charge is considered strong, and will tend to promote the S$_N$2 mechanism. Conversely, a nucleophile that does not contain an atom with a full negative charge is usually considered a weak nucleophile, and tends to promote the S$_N$1 mechanism.* There are certainly exceptions, but such a discussion is not appropriate here.

The E1 and E2 mechanisms mirror the S$_N$1 and S$_N$2 mechanisms, respectively, as they pertain to the strength of the attacking species—the base. In the E2 mechanism, the base is viewed as forcing the action. Therefore, *the stronger the base, the faster the E2 reaction.* Notice, also, that this is supported by the fact that the base does appear in the rate equation (Equation 8-4). In the E1 mechanism, however, the base does not come into play until after the rate determining step has occurred. Consequently, *the strength of the base does not affect the rate of the E1 mechanism.* Moreover, it does not show up in the rate equation (Equation 8-5).

Similar to substitution reactions, a strong base will favor the E2 reaction, whereas a weak base will favor the E1 reaction. As before, a strong base will promote the E2 mechanisms by speeding it up. A weak base, on the other hand, can't force off the leaving group and must wait until the leaving group comes off by itself.

Just as with nucleophiles, we must recognize strong and weak bases with regard to elimination reactions. Again, there is no clear cutoff. However, *as a rule of thumb, we can say that bases that are as strong or stronger than OH^- are considered strong bases, and tend to promote the E2 mechanism. Conversely, bases that are weaker than OH^- tend to promote the E1 mechanism.*

8.4 FACTOR #2: CONCENTRATION OF NUCLEOPHILE/BASE

After having derived the rate equations for the S_N1, S_N2, E1, and E2 reactions, the effect of nucleophile/base concentration on the reaction rate of each reaction is straightforward. According to Equations 8-2 and 8-3, the rate of the S_N2 reaction depends upon the concentration of the nucleophile, whereas the S_N1 reaction does not. Therefore, *increasing the concentration of the nucleophile will increase the rate of the S_N2 reaction but will have essentially no effect on the rate of the S_N1 reaction.* Similarly, according to Equations 8-4 and 8-5, the E2 reaction rate depends on the concentration of the base but the E1 reaction rate does not. Consequently, *increasing the concentration of the base will significantly increase the rate of the E2 reaction but not the rate of the E1 reaction.*

One of the lessons here is that, in general, a high concentration of a nucleophile will favor the S_N2 mechanism, and a low concentration will favor the S_N1. This is because the rate of S_N1 is not terribly affected by that concentration, whereas the rate of the S_N2 is. So, a high concentration of the nucleophile will allow the S_N2 reaction to proceed quickly—faster than the S_N1. A low concentration, on the other hand, will force the S_N2 to proceed slowly, even more slowly than the S_N1.

We have a very similar story with E1 and E2. That is, a high concentration of base tends to favor E2, and a low concentration of base tends to favor E1.

We must be careful, however, about making such statements regarding weak nucleophiles and weak bases. A nucleophile is said to be weak if it is incapable of forcing off the leaving group in an S_N2 reaction—instead, it waits for the leaving group to leave on its own, before attacking the highly reactive carbocation that is produced. A high concentration of such a weak nucleophile does nothing to allow those nucleophiles to force off the leaving group. All that would be different is the number of nucleophiles waiting for the leaving group to come off on its own. In summary, then, a weak nucleophile tends to favor the S_N1 mechanism, regardless of the concentration of that nucleophile. By analogy, a weak base tends to favor the E1 mechanism, regardless of its concentration.

8.5 FACTOR #3: STABILITY OF THE LEAVING GROUP

A leaving group leaves in each of the four reactions. Furthermore, the step in which the leaving group leaves is the rate determining step of each reaction. Therefore, the better (more stable) the leaving group, the faster the reaction—in all four cases. However, the stability of the leaving group affects the reaction rates differently. As it turns out, *the rates of the S_N1 and E1 mechanisms are significantly more sensitive to the stability of the leaving group than the S_N2 and the E2 mechanisms.* Specifically, increasing the stability of the leaving group would increase the rate of the S_N1 reaction more than it would the S_N2 reaction, and it would increase the rate of the E1 reaction more than it would the E2 reaction. This is because in the rate determining steps of the S_N1 and the E1 mechanisms, the only thing that happens is the departure of the leaving group. There is no other species available to assist the leaving group to leave. On the other hand, in the S_N2 mechanism, the nucleophile is viewed as forcing the action. It therefore provides significant assistance in the leaving of the leaving group. Similarly, the base in the E2 mechanism provides significant assistance in the leaving of the leaving group, which diminishes the importance of the leaving group stability.

The lesson here is that very good leaving groups tend to favor S_N1 over S_N2 and tend to favor E1 over E2. This is because, as a result of the high sensitivity toward the leaving group ability, a more stable leaving group speeds up the S_N1 and E1 mechanisms much more than the S_N2 and E2 mechanisms.

The cutoff between what is considered a good leaving group and a bad leaving group is somewhat fuzzy, but we can provide some benchmarks. Br^- is generally considered a very good leaving group—the Br atom, because of its large size and relatively high electronegativity, accommodates the negative charge quite well. Therefore, any leaving group that is at least as stable as Br^- should also be considered a good leaving group, favoring the S_N1 and E1 mechanisms over the S_N2 and E2 mechanisms.

Somewhat less stable is the Cl^- ion, because it is significantly smaller in size than the Br^- ion. It is, however, considered a decent leaving—decent enough that substrates containing such a leaving group can undergo S_N1 and E1 reactions. On the other hand, it is not a good enough leaving group to say that it overwhelmingly tends to favor S_N1 and E1 over S_N2 and E2. Instead, we will say that it tends to favor all four mechanisms roughly the same.

The F^- ion is less stable than Cl^-, again because of its smaller size. Consequently, substrates containing F^- do not undergo S_N1 and E1 reactions, meaning that F^- should be considered a bad leaving group. Therefore, if a substrate containing F^- as the leaving group is to undergo any of the four mechanisms, it must be either S_N2 or E2—the F^- leaving group must be forced off by a strong nucleophile. However, F^- tends to be a bad enough leaving group that such S_N2 and E2 reactions are rare.

Finally, OH^- is a terrible leaving group. The O atom cannot accommodate the negative charge as well as F because O has a smaller electronegativity. Because of that instability, we never see alcohols, R—OH, undergo S_N1 or E1 mechanisms without modification of the leaving group (discussed in Section 8.5a)—the leaving group would otherwise come off as OH^-. Moreover, OH^- is such a bad leaving group that we almost never see alcohols undergo S_N2 or E2 mechanisms.

The OH^- ion is therefore our benchmark for bad leaving groups. Any leaving group that would come off as a species as stable or less stable than OH^- is also considered a bad leaving group. Examples include CH_3O^-, H_2N^-, CH_3^-, and H^-. Any substrate containing only one of these leaving groups tends to be unreactive toward any of the four mechanisms in this chapter.

8.5a Making a Good Leaving Group Out of a Bad One

As just discussed, OH^- is a bad leaving group because the O atom does not accommodate the negative charge very well. However, it is rather straightforward to convert the HO^- leaving group into a good one. One method is to protonate the alcohol at the O atom, using a somewhat strong acid like H_2SO_4 or H_3PO_4. This would result in $R—OH_2^+$, which enables the leaving group to come off in the form of OH_2 (i.e., H_2O), or water. Since water is quite a stable molecule, bearing no formal charge, it is an *excellent* leaving group! With such a good leaving group, both the S_N1 and the E1 mechanisms become quite favorable.

Other ways of converting the bad HO^- leaving group into a good one will not be presented here, but they are all similar in that the leaving groups that are generated are very stable in the form in which they leave. Notice that in protonating the O atom before it comes off, the leaving group is stabilized by having removed the formal negative charge ($HO^- \rightarrow H_2O$). Something else that is typically done is to convert the HO^- leaving group into one that allows the formal negative charge to remain but has the negative charge involved in extensive resonance—something that we saw in Chapter 5 that can stabilize a charged species significantly.

Amines, $R—NH_2$, present a situation similar to alcohols. As is, amines are essentially unreactive toward substitution and elimination reactions, because the leaving group that would come off is H_2N^-, which is even more unstable than HO^-. However, because $R–NH_2$ is mildly basic at the N atom, acidic conditions will result in $R—NH_3^+$, so that the leaving group that can come off would be NH_3—quite a stable molecule.

8.6 FACTOR #4: NUMBER OF ALKYL SUBSTITUENTS IN THE SUBSTRATE

The number of alkyl groups (Rs) on the C bearing the L can range from zero to three. If there are no R groups and three Hs, then the C is part of a *methyl* (CH_3) group. If there is one R group, it is a **primary** (1°) **C atom**. Two and three R groups give rise to **secondary** (2°) and **tertiary** (3°) C atoms, respectively.

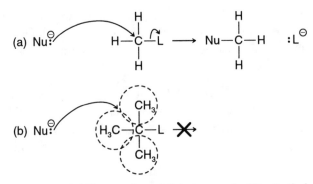

Figure 8-3 The S$_N$2 reaction with different numbers of alkyl groups on the C bearing the leaving group. (a) The H atoms are small, allowing the nucleophile to attack the C atom. (b) The R groups are bulky, providing steric hindrance that prevents the nucleophile from being able to attack the C.

The rate of the S$_N$2 reaction depends on whether the L is on a methyl, primary, secondary or tertiary C atom. The reason is that the S$_N$2 reaction rate is dictated by the nucleophile's ability to force the action, and in order to do so, it must form a bond with that C. As it turns out, *the more alkyl groups on the C bearing L, the slower the S$_N$2 rate.* Those additional R groups are bulky, making it more difficult for the nucleophile to approach the C atom and to actually form a bond with it (Figure 8-3). This phenomenon is called **steric hindrance** by the R groups. Therefore, a substrate that has the L on a tertiary C atom would slow down the S$_N$2 reaction rate significantly. On the other hand, if the C atom is either a methyl or primary carbon atom, steric hindrance is not a problem, and the S$_N$2 reaction rate is not slowed down significantly. In other words, methyl and primary C atoms favor the S$_N$2 mechanism, whereas tertiary C atoms disfavor it. Secondary C atoms are somewhat neutral in these regards.

In the E2 reaction, on the other hand, the number of R groups on the C with the leaving group has little influence on the rate of the overall reaction. This is because the base does not attack at that C atom. In fact, it does not attack any C atom—it attacks the H atom on the C atom next door, and that H atom is exposed, ready to be picked off by a base (Figure 8-4). The number of R groups on the neighboring C (bearing the L) has little impact on the exposure

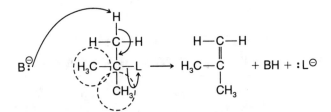

Figure 8-4 Steric hindrance has little effect in an E2 mechanism, because the H atom that is attacked is well exposed.

of the H atom. Therefore, E2 reactions can proceed readily with all three types of C atoms—primary, secondary and tertiary.

The S_N1 and E1 mechanisms share exactly the same rate determining step—the L leaves spontaneously, leaving behind a carbocation, C^+. As we saw in Section 5.9, additional R groups on that C atom help to stabilize the positive charge. The happier the resulting carbocation, the faster that rate determining step and the faster the overall reaction. Consequently, *tertiary C atoms tend to promote the S_N1 and E1 mechanisms, whereas methyl and primary C atoms tend to slow them down significantly.* Secondary C atoms are somewhat neutral in these regards, as they were with the S_N2 and E2 mechanisms.

Putting it all together, then, we see that CH_3 and primary C atoms tend to favor S_N2, whereas tertiary C atoms tend to favor S_N1 and E1. The rate of an E2 reaction, on the other hand, is not affected greatly by the number of alkyl groups on the C with the L.

8.7 FACTOR #5: SOLVENT EFFECTS

The choice of solvent can have a dramatic effect on the outcome of S_N1, S_N2, E1, and E2 reactions. The two types of solvent important in these reactions are protic and aprotic solvents, which we introduced in Chapter 7. A **protic solvent** contains a hydrogen bond donor, whereas an **aprotic solvent** does not. Examples of common protic solvents include water and alcohols (ROH). Common aprotic solvents include dimethyl sulfoxide (DMSO), dimethyl formamide (DMF), and acetone, previously shown in Figure 7-18.

Solvent can affect the outcome of nucleophilic substitution and elimination reactions by altering their reaction rates differently. The different effects on reaction rate originate from altering the strength of the nucleophile or base differently. A species that is a strong nucleophile or base has significant concentration of negative charge centered on a single atom. The large concentration of positive charge on the H atom of protic solvents enables those solvent molecules to stick quite strongly to strong nucleophiles and bases. As discussed in Chapter 7, this dramatically weakens the strength of the nucleophile, because the concentration of negative charge is tied up with the numerous partial positive charges on the H atoms of the solvent molecules (Figure 7-17). Similarly, a strong base will be significantly weakened in a protic solvent.

Recall that weak nucleophiles and weak bases tend to favor S_N1 and E1 mechanisms, respectively. As a result, *conditions that normally favor an S_N2 or E2 reaction—strong nucleophile/base—can be altered to favor an S_N1 or E1 reaction by choosing the solvent to be protic, which effectively weakens the nucleophile/base* (we will see examples of this later). On the other hand, aprotic solvents do not stick to anions as well, because they do not have a large partial positive charge on an H atom. Therefore, aprotic solvent molecules do not substantially weaken strong nucleophiles and bases.

The lesson: Aprotic solvents tend to favor S_N2 and E2, whereas protic solvents tend to favor S_N1 and E1.

8.8 SUBSTITUTION VS. ELIMINATION

Conditions that favor S_N2 reactions generally favor E2 reactions as well. This is because the same concentration of negative charge that makes a nucleophile a nucleophile also makes a base a base (as discussed in the introduction to this chapter). In other words, strong nucleophiles are usually strong bases; examples include HO^-, CH_3O^-, H_2N^-, and H_3C^-. Therefore, if HO^- reacts with a primary alkyl halide, such as CH_3CH_2Br, in an aprotic solvent, we would expect that the S_N2 mechanism proceeds at a substantial rate. This is because the nucleophile is rather strong, forcing the leaving group to leave. On the other hand, under the same conditions, we should also expect the E2 mechanism to proceed at a substantial rate. This is because the base is quite strong, which forces the removal of a proton and the leaving of the leaving group. The result is a mixture of predominantly the S_N2 and E2 products.

If the nucleophile/base is chosen wisely, however, it is possible to promote the S_N2 reaction over the E2, and vice versa. This can be done because *the rate determining step of the S_N2 reaction is different from that of the E2 reaction.* Strong nucleophilic character of the species that attacks the substrate will enhance the rate of the S_N2 reaction, while strong basic character will enhance the rate of the E2 mechanism. Therefore, in order to promote the S_N2 mechanism over the E2 mechanism, we must simply choose an attacking species that is a strong nucleophile but a weak base. Examples include Cl^-, Br^-, HS^-, $N{\equiv}C^-$, and N_3^-. Notice that they all have formal negative charges, characterizing them as strong nucleophiles. (Using the concept of charge stability you learned in Chapter 5, can you explain why these are considered weak bases?)

Conversely, in order to promote the E2 mechanism over that of the S_N2, we must find an attacking species that is a strong base but is simultaneously a weak nucleophile. A good example is the *t*-butoxide anion, $(CH_3)_3CO^-$. This species is a strong base because the negative charge is localized on an O atom and should therefore have a base strength similar to that of HO^- (in fact, because of the inductive effects of the alkyl groups, this base is even stronger than HO^-). However, it is a poor nucleophile, despite the negative charge. The reason is that the entire species is bulky, with the three methyl groups surrounding the nucleophilic O atom. Those methyl groups introduce steric hindrance that makes it difficult for the nucleophilic O atom to get close to the C atom of the substrate (Figure 8-5)—something that is otherwise necessary to form a C—O bond and to force the leaving group to leave.

Just as S_N2 and E2 reactions are favored under very similar reaction conditions, S_N1 and E1 reactions are also favored under similar reaction conditions. Both the S_N1 and E1 reactions

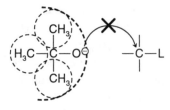

Figure 8-5 The *t*-butoxide anion is a bad nucleophile because of the steric hindrance introduced by the alkyl groups.

are favored when the substrate has L on a tertiary C atom, when L is a good leaving group, when the solvent is protic, and when the attacking species (the base or nucleophile) is weak. As it turns out, weak nucleophiles are often weak bases as well—both typically have low concentrations of negative charge on a single atom (recall arguments from Chapter 5).

Unlike S_N2 versus E2 reactions, it is very difficult to manipulate the percentage of S_N1 product compared to that of E1. The reason is that *both the S_N1 and E1 reactions share exactly the same rate determining step* (this was not the case for S_N2 versus E2). As a result, just about anything that can speed up the S_N1 mechanism will also speed up the E1 mechanism. Therefore, with S_N1 and E1 mechanisms it is very difficult to choose reaction conditions that favor one over the other. In general, S_N1 products will always be contaminated with a significant percentage of E1 products and vice versa.

8.8a Heat

Heating the reaction tends to increase the percentage of elimination products relative to substitution. The major reason for this involves **entropy,** which we discussed briefly in Chapter 7, and which you should have encountered in general chemistry in the discussion of Gibbs free energy. Entropy can be considered as *a measure of disorder*. This is important because, all else being equal, a more disordered (i.e., higher entropy) system is more highly favored.

When we compare substitution versus elimination, we can examine the amount of disorder in the products to determine the roll that entropy plays. Compare the general substitution and elimination reactions (Equations 8-6a and 8-6b), where a negatively charged species (acting either as a nucleophile or as a base) reacts with a substrate containing a leaving group. If the result is substitution, then there are two product species formed (Equation 8-6a). If the result is elimination, then there are three product species formed (Equation 8-6b). Consequently, *there is more disorder in the products of an elimination reaction than in the products of a substitution reaction*. Since competing substitution and elimination reactions start with exactly the same reactant molecules, we can say that *entropy favors elimination over substitution*.

$$\text{Nu:}^- + \text{R—L} \rightarrow \text{Nu—R} + \text{:L}^- \qquad (8.6a)$$

$$\text{B:}^- + \text{H—C—C—L} \rightarrow \text{BH} + \text{C}=\text{C} + \text{:L}^- \qquad (8.6b)$$

We have yet to answer the question of how temperature plays a role in the outcome of substitution versus elimination reactions. Much of it has to do with the fact that *higher temperatures favor disorder*. We see this when heat is added to ice, thereby melting the ice and forming water. Additional heat converts the water to steam. There is a dramatic difference in entropy (disorder) between ice (a solid, in which all of the molecules are quite well ordered to make a crystal) and steam (a gas, where the individual molecules move around independently of one another). Steam has a significantly greater entropy than ice. And, as we all know, high temperatures favor steam over ice.

Relating this to chemical reactions, all else being equal, *increasing the temperature tends to favor the reaction that produces products with higher entropy*. In this case, increasing the temperature of substitution/elimination reactions tends to favor elimination, because its products have higher entropy—there are more of them.

There are several ways to indicate that a reaction is taking place at high temperatures. One way is simply to provide the temperature at which the reaction is occurring. For example, a reaction arrow might be written $\xrightarrow[150°]{}$ to indicate that the reaction is being heated to a temperature significantly above room temperature. Another way the arrow could be written is $\xrightarrow[heat]{}$ or $\xrightarrow[\Delta]{}$ to explicitly show that heat is being added to the reaction. Although these reaction arrows may not automatically mean that elimination is occurring more prevalently than substitution, the fact that heat is being added to the reaction should suggest that elimination is a strong likelihood.

8.9 SAMPLE PROBLEMS—PUTTING IT ALL TOGETHER

As was mentioned at the beginning of this chapter, for many students, the most difficult thing to do is predict the products of a reaction that can proceed by S$_N$1, S$_N$2, E1, or E2. Even though we just went through all of the individual factors that dictate the outcome, predicting the products can still be challenging. This is because there are several pieces of information that must be considered at the same time in order to make the correct prediction. However, there is a rather straightforward way to make sense of all the information in this chapter. Getting comfortable with it requires practice, which is why this section walks you through applying what you have learned toward a variety of different situations.

To solve a problem that asks you to predict the products of a reaction that can go by any of the four reactions, consider each of the five major factors separately and then examine your results. The idea is simple: *Each of the five factors will be given roughly equal weighting in governing the outcome of the reaction, and the mechanism that the majority of the factors ends up favoring is the one that we will go with*. There may be some exceptions, but those will be left to your traditional textbook.

In the first of several problems, you are asked to predict the products of the following reaction, including stereochemistry if appropriate.

$$CH_3CH_2CH_2Cl + Br^- \xrightarrow[DMSO]{} \text{?}$$

Notice that there are two things that we are asked for—the products (i.e., the connectivity) and stereochemistry. Realize that both of these fall straight out of the mechanism. That is, if we can predict the mechanism that will occur, both the products and the stereochemistry will be determined. To predict the mechanism, we must apply what we know about the five factors.

The first factor we looked at was the strength of the nucleophile/base. In this case, Br^- is the nucleophile, and because it bears a full negative charge, it is considered a strong nucleophile. The fact that it is a strong nucleophile suggests that it has the ability to force the action in a substitution reaction, thereby favoring the S_N2 mechanism over the S_N1. We keep track by entering this result into Table 8-1.

What about elimination? As we discussed earlier, we must determine whether Br^- is considered a strong or a weak base, and to do so, we must compare the base strength of Br^- to that of OH^-. Because Br can better accommodate the negative charge (as a result of its larger size than O), Br^- is a much weaker base than OH^-, and we consider Br^- to be a weak base. As a result, Br^- will not have a tendency to force the action in an elimination reaction. Instead, it will sit back and wait for the leaving group to come off before it acts as a base to pull off the adjacent proton. That is, it will tend to favor an E1 over an E2. Again, we enter this information into Table 8-1.

The next factor that we looked at was the concentration of the nucleophile/base. With the reaction written the way it is, we can assume that both the Br^- and the substrate are present in significant amounts. If this were *not* the case, it would have been made explicit to us that one of the reactants was dilute. For example, "dil. Br^-" may be written above the reaction arrow. Assuming that Br^- is fairly concentrated in our solution, S_N2 will be favored over

Table 8-1

Factor	S_N1	S_N2	E1	E2
Strength		√	√	
Concentration		√		
Leaving Group	√	√	√	√
R Groups		√		√
Solvent		√		√
Total	1	5	2	3

S_N1. Again, this is because Br^- is viewed as forcing the action, and the greater the concentration of this nucleophile, the more it will do so.

On the other hand, if we view Br^- as a base, then the fact that it is present in large amounts might suggest that E2 is favored over E1. However, we identified Br^- as a weak base, meaning that it does not force the action in an E2. The large concentration of a weak base does not change the situation. Therefore, we may ignore this factor in considering elimination reactions.

The third factor we examined was stability of the leaving group. In this case, the leaving group comes off in the form of Cl^-. To determine which mechanism(s) this factor favors, we must determine where Cl^- falls on the scale of good versus bad leaving groups. To do so, we must compare the stability of Cl^- with that of Br^- and with that of OH^-. Cl^- is more stable than OH^- because the Cl atom is larger than the O. Therefore, Cl^- is not considered a bad leaving group. On the other hand, Cl^- is less stable than Br^-, so that it is not considered a very good leaving group either. Overall, then, Cl^- is a decent leaving group, and therefore tends to favor all four mechanisms roughly equally. We enter these results into the table by placing a check under all four mechanisms (alternatively, we could have left all of the entries blank).

The next factor to take into account is the number of alkyl substituents on the C atom containing the leaving group. In this case it is one—the carbon containing the leaving group is a primary C. This will therefore tend to favor S_N2 over S_N1, for two reasons. First, the absence of steric effects allows the nucleophile to approach the C from the opposite side of the leaving group and to force the action. Second, the rate determining step of the S_N1 mechanism would be heavily disfavored and would therefore be quite slow, because the product of that step would be an unstable primary carbocation. For the same reason, the rate of the E1 mechanism would be quite slow, suggesting that the E2 would be favored over the E1.

The final factor that we examined was solvent effects. From the information that we are given, it appears that DMSO is the solvent, which is an aprotic solvent. Aprotic solvents tend to favor S_N2 and E2 mechanisms, because they do not weaken the strength of the nuclephile/base.

Now that we have considered all five factors, let's tally them. This is done in Table 8-1. The S_N2 mechanism appears to be the most heavily favored—it is favored by all of the five factors. The E2 mechanism is next, with three factors favoring it, followed by two factors for E1. The S_N1 mechanism is the least favored, with only one. We would therefore predict that the reaction we were presented with would proceed via the S_N2 mechanism faster than any of the others, and the S_N2 products would be the major products.

Recall that we said, in general, if reaction conditions favor S_N2, they also favor E2. However, that does not appear to be the case with this example. Clearly, from the table, it appears that the S_N2 reaction is favored over the E2. Why is this? It primarily has to do with the

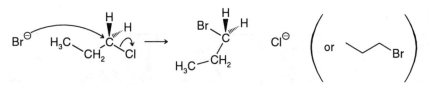

Figure 8-6 S$_N$2 mechanism of the reaction between Br$^-$ and H$_3$CCH$_2$CH$_2$Cl.

strength of Br$^-$ as a nucleophile compared to its strength as a base. We argued that it is a rather strong nucleophile but a weak base. It is therefore able to force the action in an S$_N$2 mechanism but not in an E2, allowing it to discern between the two mechanisms.

Finally, let's predict the products, taking into account stereochemistry. We do so using the S$_N$2 mechanism, which is found in Figure 8-6. The Br$^-$ attacks the C from the opposite side of the leaving group, and the leaving group leaves at the same time. Furthermore, the three other substituents—the two H atoms and the R group—are inverted in the process. Notice that the product is achiral (as was the reactant substrate), meaning that there is no stereochemistry that must be shown explicitly.

Let's change the problem slightly, so that the substrate is that given in the reaction here. All the arguments pertaining to the five factors remain the same, except for the number of R groups. Here, the C with the leaving group is bonded to two R groups, making it a secondary C atom. The additional steric hindrance at the C atom slows the rate of the S$_N$2 reaction. It does not affect the rate of the E2 mechanism significantly, because the proton that is removed is on the C atom next door. The S$_N$1 and E1 reactions, on the other hand, are sped up a bit, because the carbocation that results from the leaving of the leaving group is a secondary carbocation, as opposed to a primary. Consequently, a secondary C atom does not heavily favor any one of the four reactions over another—it favors them all roughly equally. In fact, under certain overall conditions of the remaining four factors, substrates in which the leaving group is on a secondary C can lead to a mixture of products from all four reactions.

$$Br^{\ominus} + H_3C\!-\!\!\overset{\displaystyle CH_3}{\underset{\displaystyle CH_2}{\overset{|}{\underset{|}{C}}}}\!-\!Cl \xrightarrow{DMSO} ?$$

Given the new substrate, the table changes slightly (Table 8-2). Notice that checks were placed at all entries in the row for the number of R groups, since we argued that with a secondary C atom that contains the leaving group, all four mechanisms are favored roughly the same. The columns are retotaled, and it still appears that the S$_N$2 mechanism wins out over the others.

Once again, finishing the problem entails using the S$_N$2 mechanism (Figure 8-7) to predict the products with the correct stereochemistry. This time, the reactant substrate is chiral,

Table 8-2

Factor	S$_N$1	S$_N$2	E1	E2
Strength		√		
Concentration		√		
Leaving Group	√	√	√	√
R Groups	√	√	√	√
Solvent		√		√
Total	2	5	2	3

and the stereocenter is the C atom bearing the leaving group. The Br⁻ again attacks the C atom from the opposite side of the leaving group, the leaving group comes off, and the remaining two R groups and H atom flip over to the other side. Notice that the product in this reaction is chiral, containing one stereocenter. The correct three-dimensional representation, and therefore stereochemistry, follows directly from the mechanism and is as shown. Notice also that because the reactant substrate was chiral, and the reaction proceeded in one step, the product is NOT a racemic mixture of enantiomers.

In the next problem, let's predict the products of the following reaction, which denotes that the substrate is dissolved in water.

$$(CH_3)_3CBr \xrightarrow{H_2O} ?$$

As before, we analyze the effects of each of the five factors on the four mechanisms. The first factor is the strength of the nucleophile/base. But what is the nucleophile/base? Clearly there is a substrate—the species being dissolved that contains Br as the leaving group. That must mean that the H$_2$O is the nucleophile/base, such that the solvent is behaving as a reactant (this brand of reactions is called **solvolysis**). Since water is a weak nucleophile and a weak base, this factor favors S$_N$1 and E1 over S$_N$2 and E2, respectively. These results are entered into Table 8-3.

The second factor, concentration of the nucleophile/base, is irrelevant in this reaction. There is a large concentration of water, but as we saw previously, a large concentration of a

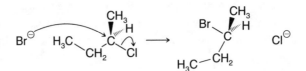

Figure 8-7 S$_N$2 mechanism for the reaction between Br⁻ and CH$_3$CH$_2$CH(CH$_3$)Cl. The product is chiral, and the configuration about the stereocenter is the one shown.

Table 8-3

Factor	S$_N$1	S$_N$2	E1	E2
Strength	√		√	
Concentration				
Leaving Group	√		√	
R Groups	√		√	
Solvent	√		√	
Total	4	0	4	0

weak nucleophile/base does not enable that nucleophile/base to force the action. Therefore, no checks are entered in for this row.

The third factor, stability of the leaving group, favors S$_N$1 and E1. The leaving group comes off as Br⁻, which, as was mentioned previously, is a benchmark for a good leaving group.

The fourth factor, number of R groups on the C bonded to the leaving group, favors S$_N$1 and E1 over S$_N$2 and E2. This is because it is a tertiary C atom, as it is bonded to three methyl groups. Those three methyl groups provide significant steric hindrance to the attack of the nucleophile in an S$_N$2 mechanism (and, to a lesser extent, the attack of the base in an E2). At the same time, because of their electron-donating capability, they provide significant stability to the resulting carbocation—a tertiary carbocation—which should speed up the rate determining step of the S$_N$1 and E1 mechanisms.

The final factor is the solvent. Water is a protic solvent and therefore favors S$_N$1 and E1 over S$_N$2 and E2.

The overall tally of the five factors is given in Table 8-3. The results are unanimously in the favor of S$_N$1 and E1 over the other two reactions. Since we are given no other information, the best answer is that the result is a mixture of products from the S$_N$1 and E1 reactions. It makes sense that this is the case, since we argued earlier that if S$_N$1 is favored, we should expect a significant amount of E1, and vice versa. This is because the two mechanisms share exactly the same rate determining step.

Plugging the specific molecules into the appropriate S$_N$1 and E1 mechanisms yields the mechanisms shown in Figure 8-8. The products are then a mixture of the following two compounds:

$$H_3C-\underset{\underset{CH_3}{|}}{\overset{\overset{CH_3}{|}}{C}}-\underset{H}{\overset{}{O}} \quad + \quad H_3C-\underset{\underset{CH_3}{|}}{\overset{\overset{CH_2}{||}}{C}}$$

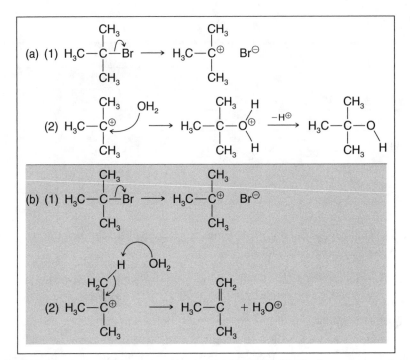

Figure 8-8 (a) S_N1 mechanism for the reaction between H_2O and $(CH_3)_3CBr$. The first two steps only are technically the S_N1 mechanism. The third step is simply a deprotonation to achieve a neutral molecule. (b) E1 mechanism between the same two molecules.

In yet another reaction, let's predict the product of the following reaction:

OH

$+ 85\% H_3PO_4 \xrightarrow{\Delta}$?

This particular problem is a bit different from the previous ones in that it is not immediately obvious which molecule is the substrate and which one is the nucleophile/base. It turns out that as it is written, there is no molecule that has any kind of decent leaving group. If the compound on the left were to act as the substrate, then the leaving group would have to be HO^-, which, as we saw before, is a terrible leaving group. However, the OH group can be protonated by the phosphoric acid to make OH_2^+ (recall Section 8.5a), which can come off in the form of H_2O—what we identified to be a very good leaving group.

The nucleophile/base turns out to be water. The acid solution is only 85 percent H_3PO_4, meaning that the remaining 15 percent of the solution is water. The water can act as a weak base or as a weak nucleophile.

We are now in a position to tabulate which reactions each of the five factors favors (Table 8-4). The first factor, strength of the nucleophile/base (H_2O), favors both E1 and S_N1, as we saw

Table 8-4

Factor	S$_N$1	S$_N$2	E1	E2
Strength	√		√	
Concentration				
Leaving Group	√		√	
R Groups	√	√	√	√
Solvent	√		√	
Total	4	1	4	1

before. The second factor is irrelevant once again, because the nucleophile/base strength is weak. The leaving group (H$_2$O) is very stable and favors S$_N$1 and E1. The C bonded to the leaving group is a secondary C and therefore favors all mechanisms roughly equally. And finally, the solvent can be water or the alcohol reactant itself (both are present in significant amounts), both of which are protic and favor S$_N$1 and E1. At this point, Table 8-4 suggests that S$_N$1 and E1 are both favored, at four factors apiece.

It appears that, as before, we are going to be stuck with a mixture of both S$_N$1 and E1 as the major products. There is one difference, though. In this reaction, we are explicitly told that the reaction mixture is heated (signified by Δ), which is a signal that the balance is tipped in favor of elimination over substitution. Therefore, the predominant mechanism of the four should be E1, and the major products are obtained by plugging in the specific reactants, shown in Figure 8-9. Consequently, the **major product** is expected to be cyclohexene (C$_6$H$_{10}$), shown here. Because the five factors also favor

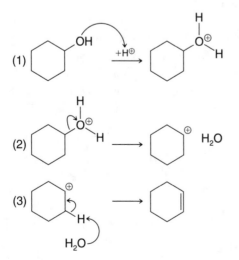

Figure 8-9 E1 mechanism for the reaction between C$_6$H$_{11}$OH and 85 percent H$_3$PO$_4$. Technically, steps 2 and 3 are the two steps of the E1 mechanism, and the first step converts a bad leaving group (OH$^-$) into a good leaving group (H$_2$O).

Table 8-5

Factor	S$_N$1	S$_N$2	E1	E2
Strength	√			√
Concentration				√
Leaving Group	√	√	√	√
R Groups	√	√	√	√
Solvent		√		√
Total	3	3	2	5

S$_N$1, we should expect significant contamination of the S$_N$1 product, called the **minor product.**

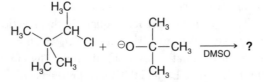

Let's examine one more reaction.

(CH$_3$)$_3$CO$^-$ is our example of a strong base but a weak nucleophile. This is because the negative charge on the base/nucleophile is on an O atom, similar to HO$^-$ (a strong base), but its bulkiness prevents the O atom from forming a bond with the C atom of the substrate. The fact that it is a weak nucleophile favors S$_N$1, and the fact that it is a strong base favors E2. Next, the concentration of the base is high, suggesting E2; although the concentration of the nucleophile is also high, it is not considered, since it is a weak nucleophile. The leaving group (Cl$^-$) is one that favors all four mechanisms roughly equally. The C with the leaving group is secondary, which also favors all four mechanisms roughly equally. Finally, the solvent is aprotic, suggesting S$_N$2 and E2. The results are summarized in Table 8-5, where it appears that E2 is the favored reaction. The resulting mechanism is shown in Figure 8-10.

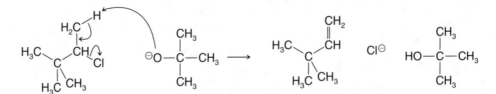

Figure 8-10 E2 mechanism for the reaction corresponding to Table 8-5.

Problems

8.1 For each reaction here, (a) predict the predominant mechanism(s); (b) draw out the detailed mechanism(s), including all curved arrows; and (c) using the mechanism, predict the major product(s), including stereochemistry where appropriate.

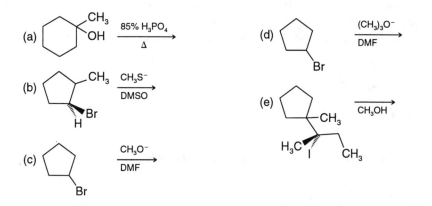

8.2 Provide the missing reagents and/or reaction conditions (e.g., solvent, heat) necessary to produce each of the following products from S_N1, S_N2, E1, or E2 reactions. Reaction conditions are placed in the box below the arrow. Pay attention to stereochemistry where appropriate.

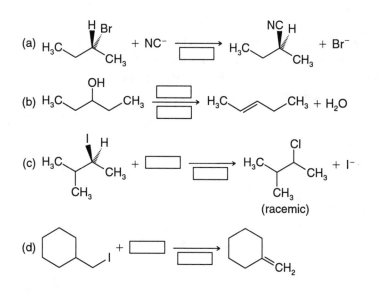

8.3 The molecule here cannot be used as a substrate for elimination. Why not?

8.4 A student attempted to carry out a substitution reaction using molecule (a) as the substrate. The resulting product was molecule (b). Which nucleophilic substitution mechanism had occurred—S$_N$1 or S$_N$2? How can you tell? (Hint: See Chapter 6.)

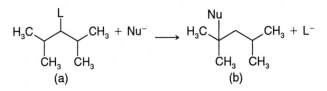

8.5 Predict the product of the following solvolysis reaction.

8.6 Predict the product of the following solvolysis reaction in which oxygen-18 labeled water is used as the solvent. (See Problem 6.7.)

8.7 In the text, we argued that substrates in which the leaving group is on a primary carbon atom tend not to react via the S$_N$1 mechanism. However, benzyl substrates (shown here) often undergo S$_N$1. Explain why. (Hint: Write out the complete S$_N$1 mechanism involving the benzyl substrate and a nucleophile.)

Concluding Remarks: What Now?

9

Students from my summer organic prep course overwhelmingly feel that they are prepared to face their year-long organic chemistry course. However, they unanimously express one concern: What do I do if my professor doesn't teach like you?

Quite frankly, you should expect your course to be different, simply because of the way that most organic chemistry textbooks (and courses) are organized—that is, by functional group instead of by fundamental concept. It may be that your professor *does* focus heavily on fundamental concepts and understanding mechanisms (thank them if they do), but it may feel different simply because multiple fundamental concepts are introduced at the same time in order to explain the variety of reactions a given functional group undergoes. In other instances your course may be very different in that your professor places less emphasis on reaction mechanisms.

In either case, this does *not* mean that you should abandon what you have learned here. Instead, it means that you must assimilate your understanding of fundamental concepts we have learned with the way material is presented in your class. This book provides you the tools, and the rest is up to you. If a reaction is presented to you without a mechanism, get ahold of the mechanism, either by looking it up or by asking your professor or TA. Convince yourself that each step of the mechanism is reasonable using the concepts from this book. If even one step of a mechanism does not make sense to you, ask questions. If you don't take it upon yourself to do so expect a lot of memorization!

A second concern from my students arose after they had returned to their respective institutions. Up through the middle of the first semester, they had great success without resorting to flash cards and memorization, but the increasing number of reactions made flash cards more appealing. They essentially asked me: "What do I do now?" The following summarizes the advice I gave them, which, they tell me, helps considerably.

If you get overwhelmed by the sheer number of reactions, then you are doing something wrong. You are probably straying from focusing on the mechanism. I say this because even though you face dozens upon dozens of reactions in the first semester, and will face many more in the second semester, the number of *mechanisms* that dictate those reactions is quite small. In the first semester, the total number of mechanisms is somewhere around 10. In the second semester, there may be an additional 10 or so, and several of them are nothing more than a simple twist on, or a combination of, the mechanisms you learn in the first semester. Therefore, if you focus on the mechanisms, all of the reactions you see in your year of organic chemistry, which may be *several hundred,* are reduced to only a couple dozen mechanisms at most.

Once you understand that all of the organic chemistry reactions you encounter are governed by a handful of mechanisms, things will get a lot easier. But you still have to work. As mentioned earlier, you have to work at using the concepts you learned in this book to *understand* each mechanism. You also have to work at *using* the mechanism. Perhaps the most difficulty students have in seeing where a mechanism is useful is in problems that simply ask you to predict the products of a given reaction. But those are the types of problems where knowing and understanding mechanisms is the *most* useful! When you encounter a problem like this, write out the COMPLETE mechanism that those reactant molecules will undergo—even if the problem does not ask you to do so! That includes EVERY step of the mechanism, and EVERY curved arrow, including formal charge. All along, remind yourself why *that* particular step, and not some other one. Think about electron rich to electron poor. Think about steric hindrance and charge stability, and think about stereochemistry. When the mechanism is finished, the answer will be sitting right there on the paper. If the problem asks for stereochemistry of the product, the mechanism you just wrote out contains that information.

At first, writing out the mechanism when it is not asked for might require more time than not doing so. In the long run, however, the opposite is true—you begin to *save* time. This is especially true as you become more practiced with those mechanisms, because you are eventually able to do many of the steps in your head (which is different from skipping them entirely!). You will get faster, while maintaining your accuracy in predicting the products.

So where does this end up saving you time? First, by becoming well practiced with the mechanism, you will be well prepared to answer any question about that mechanism on the exam, especially one that presents an overall reaction and asks you for its detailed mechanism. You will therefore not need to study the reactions separately from the mechanisms, as those who primarily memorize would be forced to do. Second, instead of spending your time memorizing the hundreds of seemingly different reactions, you will only have to spend time working with a handful of mechanisms. Many reactions share exactly the same mechanism, as we showed in Chapter 6. Others might not be exactly the same mechanism but might share many of the same steps. For those mechanisms that are different by a simple "twist," all you must learn and understand is that twist.

Another great advantage to working with the mechanisms is what happens when you draw a blank on an exam. If you memorized, then all you can do is sit back and wait for divine inspiration. On the other hand, if you had primarily focused on electron rich to electron poor, you can begin to rederive the mechanism. It may be that all you need is the first step in front of you to get you back on the right track.

Most importantly, by working primarily with the mechanisms instead of memorization, you will be much less frustrated. Things will actually make sense! You will become much more accurate in your answers, while spending less time studying than your friends who memorize. With greater accuracy, and less wasted time, you will wonder why organic chemistry is the beast that everyone makes it out to be.

Good luck!

Solutions to Problems

CHAPTER 2

2.1

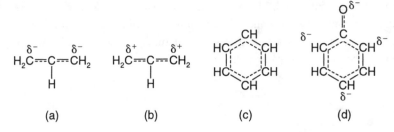

(a)   (b)   (c)   (d)

2.2 The oxygen atom has eight total electrons. Its configuration is $1s^2 2s^2 2p^4$. There are two unpaired electrons, both in 2p orbitals. There are two core electrons (the 1s electrons) and six valence electrons. The Lewis structure is below.

:Ö:

2.3

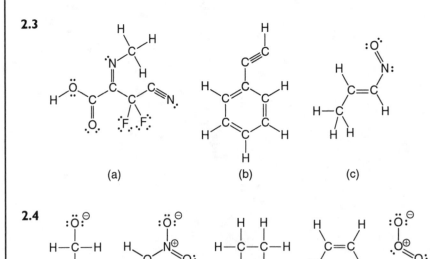

(a)   (b)   (c)

2.4

(a)   (b)   (c)   (d)   (e)

2.5

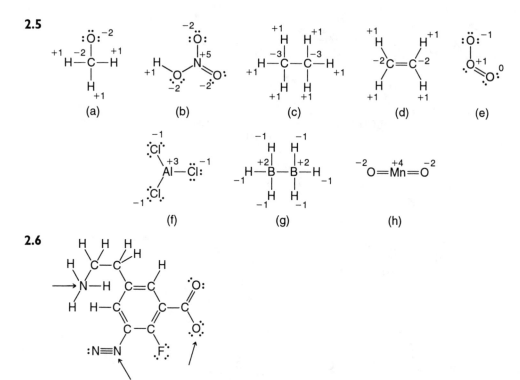

(a)   (b)   (c)   (d)   (e)

(f)   (g)   (h)

2.6

2.7

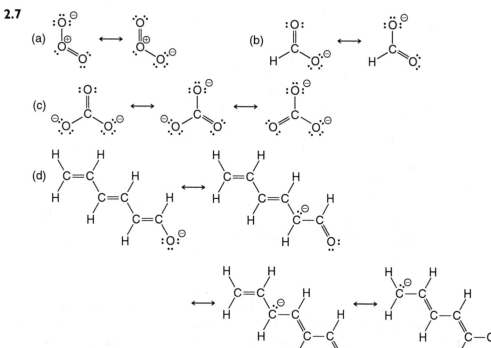

2.8

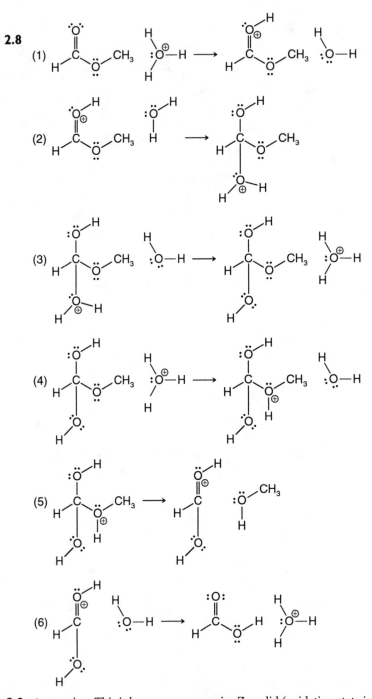

2.9 Answer is a. This is because we recognize Zn solid (oxidation state is zero) as a reducing agent, and a is the only answer showing an electron (the central C) that has been reduced.

2.10 a, b, and g. In all three cases, the atoms do not remain frozen from one structure to the other.

CHAPTER 3

3.1 (a) C and O are both sp^3. (b) C and the doubly bonded O are both sp^2. F and the singly bonded O are both sp^3. (c) The C of CH_3 is sp^3, and all other C atoms are sp^2. (d) Both triply bonded C atoms are sp. The other two C atoms and the N are all sp^3.

3.2 Answer is b. The bond is the one that attaches the triply bonded C to the C of the ring. The triply bonded C is sp-hybridized, and the ring Cs are all sp^2-hybridized.

3.3 b and c only.

3.4

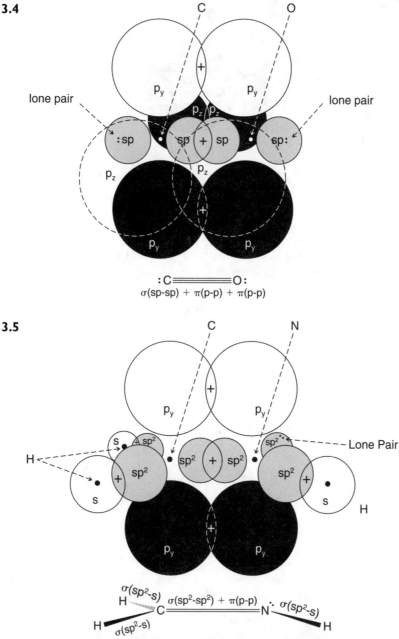

$$:C\equiv\!\equiv\!\equiv O:$$
$$\sigma(sp\text{-}sp) + \pi(p\text{-}p) + \pi(p\text{-}p)$$

3.5

3.6

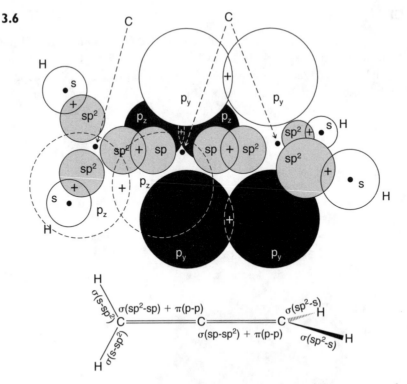

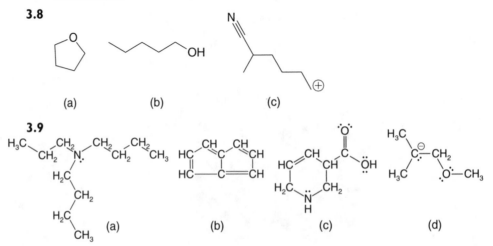

The H atoms on the left are in a plane perpendicular to the H atoms on the right. This is dictated by the fact that the p_y orbital on the central C atom is used in the π-bond on the right, and the p_z orbital is used in the π-bond on the left.

3.7 No. There is allowed rotation about the single bond connecting the middle two carbon atoms.

3.8

(a)　　　　　　(b)　　　　　　(c)

3.9

(a)　　　　　(b)　　　　　(c)　　　　　(d)

3.10 Tetrahedral, because it has 4 groups of electrons—3 single bonds and 1 lone pair.

CHAPTER 4

4.1 (a) constitutional isomers, (b) unrelated, (c) diastereomers, (d) enantiomers, (e) same molecules, (f) enantiomers, (g) diastereomers

4.2 b, d, f, g are the only chiral species.

4.3 (a) 4, (b) 5, (c) 2, (d) 0, (e) 4, (f) 2

4.4

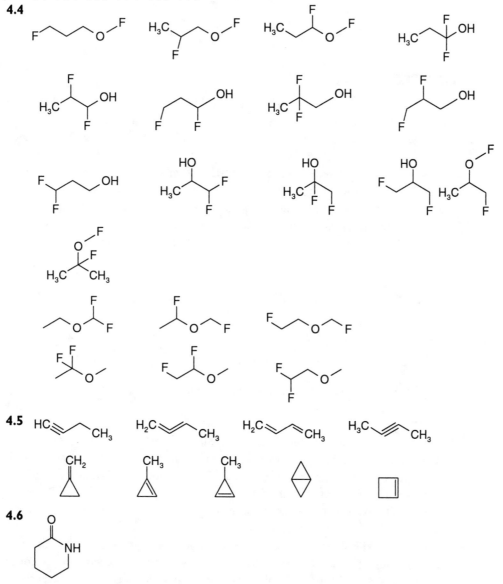

4.5

4.6

4.7 Yes, the stereocenter will be formed in equal amounts of R and S configurations. This is because the stereocenter is formed in the first step, where the reactants and reaction conditions are achiral.

4.8 The product would be formed as a racemic mixture, because the reactants are achiral, and the product is chiral with one stereocenter—the C atom bonded to three other C atoms.

4.9 The reaction is below. Notice that there is a formation of a stereocenter in the presence of a stereocenter that already exists. Therefore, that stereocenter will be formed as a mixture of R and S, but not in equal amounts.

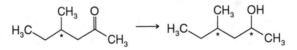

CHAPTER 5

5.1 It is tempting to say that the resonance contributor on the left is better, because the positive charge would prefer to be on a C than on an O. However, with the positive charge on the O atom, all atoms in the contributor have their octets fulfilled, making it a better contributor.

5.2 The products of protonating each of the two oxygen atoms are shown below. In the first of the two products, there is a resonance contributor that serves to allow the positive charge to be shared on the second O atom. In the second of the two products, there is no other resonance contributor, suggesting that the positive charge is stuck on just that one O atom. The first of the two products is therefore more stable and is the one that is formed.

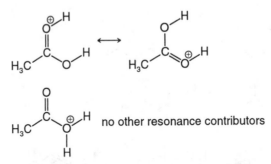

5.3 We can ignore X^- in both reactions, because they are reactants in both. We can also ignore the neutral species, taking them to be about the same stabilities. That leaves Cl^- and CH_3O^- to look at on the product side. Cl^- is a more stable anion than CH_3O^-, because Cl is below F in the periodic table (making Cl a larger atom), and F is to the right of O (giving F a higher electronegativity). The reaction forming Cl^- is more downhill, and therefore proceeds faster.

5.4 Equation 5-13 illustrates how the negatively charged nucleophiles can react, and Equation 5-14 illustrates how the uncharged nucleophiles can react. In general, the reaction in Equation 5-13 will be more downhill than the reaction in Equation 5-14, because the reactants in Equation 5-13 are less stable (extra negative charge), and the products in Equation 5-13 are more stable (no positive charge). Therefore, Cl^-, CH_3^-, HS^-, and I^- will all react faster than H_2O and H_2S. The strengths of H_2O and H_2S were compared in the text (Equations 5-17 and 5-18), and we showed that because S can accommodate a positive charge better than O, H_2S is a stronger nucleophile. Within the negatively charged nucleophiles, we examine just Equation 5-13, where a positive charge and a negative charge appear on the left side of the equation, and no charges appear on the right side. The positively charged species can be ignored because it appears in every reaction, and the products can be ignored because they are neutral. We are left to compare the stability of just the negatively charged nucleophile. Charge stability is in the order: $CH_3^- < HS^- < Cl^- < I^-$, meaning that the most downhill reaction is that with CH_3^- as the nucleophile, and the least downhill reaction is that with I^- as the nucleophile. Nucleophile strength is therefore in the order: $I^- < Cl^- < HS^- < CH_3^-$. Overall, then, the order of nucleophile strength is: $H_2O < H_2S < I^- < Cl^- < HS^- < CH_3^-$.

5.5 In all three reactions, a neutral reactant breaks apart into a cation and an anion. Comparing reactions a and c, we can ignore the reactant molecules because they are neutral and therefore have about the same stabilities. We can also ignore the cations, because they are exactly the same in both reactions. That leaves Br^- and Cl^- to compare as products. Br^- is more stable, because it is below Cl in the periodic table and is therefore larger. As a result, reaction a is more downhill and is therefore faster than reaction c. Now comparing reactions b and c, we can ignore everything except the two cations. In reaction c, the cation is more stable, because the C^+ is bonded to three alkyl groups, whereas in reaction b, it is bonded to only two. Therefore, reaction c is more downhill and should be faster than reaction b. Consequently, the order of the rates of reaction is: b < c < a.

5.6 According to arguments from the chapter, base strength increases in the order of:

$$H_2O < NH_3 < Br^- < F^- < HO^- < NH_2^- < CH_3NH^- < CH_3CH_2^-$$

If you look up the acidities experimentally determined in water, you will find a slightly different order. This has to do with effects of the solvent (Chapter 7).

5.7 f < c < a < d < g < b < e < h

Notice that f is the only one for which there is no resonance contributor to allow the negative charge to be shared between two O atoms. The rest of the order is determined by inductive effects—electron withdrawing by Cl and F atoms, and electron donating by the CH_3 group.

5.8 Using resonance arguments, HNO_3 is a stronger acid. This is because the resulting anion, NO_3^-, has a total of three resonance contributors, whereas the resulting anion from the other acid, HCO_2^-, has only two. Therefore, the negative charge is less concentrated on the O atoms in NO_3^- than it is on the O atoms in HCO_2^-. Consequently, the negative charge is more stabilized in NO_3^-.

5.9 There are a total of four resonance contributors—the one provided in the problem and the following three:

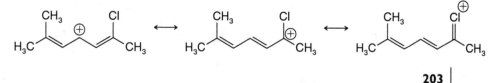

The best one is the one on the right, because it is the only one of the four that has all atoms with octets. The next best is the one given in the problem, because of the inductive electron donating effects from the three R groups. Next is the resonance structure on the left given here, because only two R groups are inductively electron donating. The worst one is the one in the middle. This is because the Cl atom is electron withdrawing compared to H atoms and therefore pulls more electron density away from the C^+. This serves to increase the positive charge already on it, making it more unhappy.

5.10 The compound with the double bonds is a stronger acid, because the anion that results from removing a proton is one that has resonance contributors—five in all. On the other hand, the anion that results from removing the proton from the compound with all single bonds is one that has no other resonance contributors.

CHAPTER 6

6.1

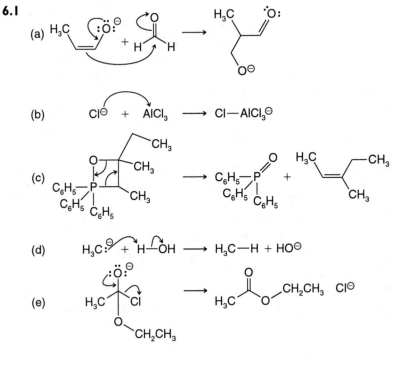

6.2

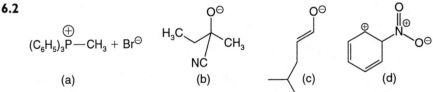

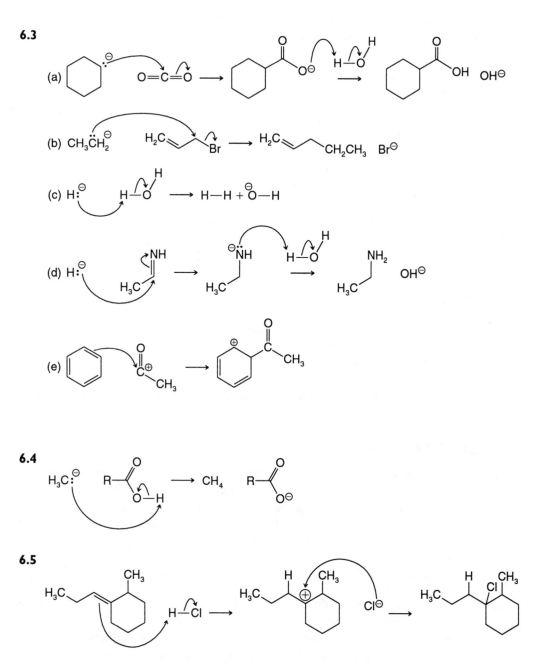

6.3

(a)

(b) CH₃ĊH₂

(c) H:⊖

(d) H:⊖

(e)

6.4

6.5

In the first step, the H could have gone to either of the doubly bonded C atoms. In the mechanism shown, the H goes to the C that results in a C⁺ with three alkyl groups surrounding it. If the H went to the other C, it would result in a C⁺ with only two alkyl groups around it. Because alkyl groups are electron donating, they stabilize the positive charge on C. Therefore, the mechanism shown is the better choice.

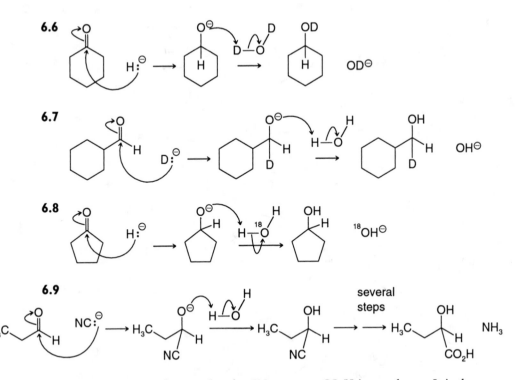

6.6

6.7

6.8

6.9

The mechanism of converting the CN group to CO_2H is not shown. It is the same mechanism provided in Figure 6-12.

CHAPTER 7

7.1 $(CH_3)_2C{=}O$ has a strong bond dipole along the $C{=}O$ bond, giving the O atom a large δ^-. Therefore, that O atom can act as a hydrogen bond acceptor, and can undergo hydrogen bonding with choices c and h—molecules with hydrogen bond donors. The remaining molecules have no hydrogen bond donors (some have hydrogen bond acceptors, but hydrogen bonds cannot form with only hydrogen bond acceptors). Therefore, c and h are the two most soluble. Of those two, c is more soluble because it has two hydrogen bond acceptors, whereas h has only one.

Because acetone has a dipole moment, it can undergo dipole-dipole interactions with other polar compounds, such as choices a, e, and g. Because C—F bonds are more polar than C—Cl bonds, a and g are more soluble than e. Furthermore, a is more soluble than g because a has two C—F bonds on the same carbon, whereas g has only one.

Choices b, d, and f are the three least soluble molecules, because they have no dipole moment. The strongest intermolecular interaction they can form with acetone is dipole–induced dipole interactions. The strongest of those interactions are with f, because it has the most electrons. Recall that in general, more electrons means a stronger induced dipole. Next is b, and finally d.

Overall, then, the order of increasing solubility is:

$$d < b < f < e < g < a < h < c$$

7.2 Because this is in DMSO, the order of nucleophile strength is the same in the situation in which there is no solvent.

a CH_3O^-

b CH_3CH_2SH

c HS^-

d comparing CH_3^- to OH^-, CH_3^- is stronger

e Cl^-

f HS^-

g H_2P^-

h $CH_3CH_2NH_2$

Note that in h, although the positively charged species is considered less stable, it is a terrible nucleophile. This is because a nucleophile is defined as something that will seek out some center with excess positive charge, and like charges repel each other.

7.3 Because this is in water, we should expect reversals only when the same negative charge is on two different atoms that are substantially different in size—that is, lower in the periodic table. We therefore expect reversals in e and g.

7.4 $e < d < c < g < a < b < f$. Species d and e are nonpolar and are therefore capable of only induced dipole–induced dipole interactions with other molecules of them-selves—those are the weakest of the intermolecular interactions we examined. The strength of interaction between molecules of e are less than that between molecules of d, because e is smaller and possesses fewer electrons. Species c, because it has a per-manent dipole, is capable of dipole-dipole interactions, which are stronger than induced dipole–induced dipole interactions. Species g, a, b, and f all are capable of hydrogen bonding with other molecules of themselves. Of those, f is capable of form-ing the most donor-acceptor pairs, followed by b. Therefore, hydrogen bonding is strongest in f and next strongest in b. g and a are tied in the department of number of donor-acceptor pairs, but the strength of a hydrogen bond with oxygen as the elec-tronegative atoms is stronger than the hydrogen bond with nitrogen as the elec-tronegative atoms. This is because oxygen is more electronegative and will generally possess a larger concentration of negative charge on it. Oxygen will further cause the hydrogen atom of the hydrogen bond donor to possess a greater concentration of positive charge than will a nitrogen atom.

7.5 This enol is stable because of the internal hydrogen bond that results. The OH is the hydrogen-bond acceptor, and the doubly bonded O atom is the hydrogen bond acceptor, as shown below.

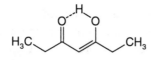

7.6 The species on the right will be weakened more. The reason is that the negative charge is localized on one oxygen atom, whereas in the species on the left, resonance allows that negative charge to be shared among both O atoms. Therefore, the negative charge in the species on the right is much more concentrated, and as a result is much better solvated. That is, the strong partial positive charge on each protic solvent molecule can stick to it much better. The extra stability from solvation in the species on the right, however, is not enough to make it a weaker nucleophile than the species on the left.

CHAPTER 8

8.1

a *E1 mechanism.* Excellent leaving group (H_2O, after the acid protonates the OH) favors E1 and S_N1. Weak nucleophile and base (H_2O) favors S_N1 and E1. Tertiary carbon bonded to leaving group favors S_N1 and E1. Protic solvent (H_2O or alcohol reactant) favors S_N1 and E1. Heat favors Elimination.

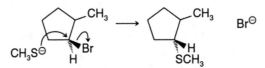

b *S_N2 mechanism.* Strong nucleophile (CH_3S^-) favors S_N2. Base (CH_3S^-), which is weaker than OH^-, favors E1. High concentration of nucleophile favors S_N2. Aprotic solvent favors S_N2 and E2. Good leaving group (Br^-) favors S_N1 and E1. Secondary C favors all four.

Notice in the mechanism that the nucleophile attacks from the end opposite the Br and therefore ends up on the opposite side.

c *Both S_N2 and E2 mechanisms.* Strong nucleophile (CH_3O^-) favors S_N2. Strong base (CH_3O^-) favors E2. High concentration of nucleophile/base favors S_N2 and E2. Secondary C atom favors all four. Aprotic solvent favors S_N2 and E2. Good leaving group (Br^-) favors S_N1 and E1.

Note, the products are achiral, so no stereochemistry must be provided.

d *E2 mechanism.* Everything is exactly the same as in (c), except that the nucleophile/base, $(CH_3)_3CO^-$, is bulky and therefore a weak nucleophile. But it remains a strong base.

e *Both S$_N$1 and E1 mechanisms.* Good leaving group (I⁻) favors S$_N$1 and E1. Weak nucleophile (HOCH$_3$) favors S$_N$1. Weak base (HOCH$_3$) favors E1. Protic solvent (HOCH$_3$) favors S$_N$1 and E1. Tertiary C favors S$_N$1 and E1. The S$_N$1 and E1 mechanisms are shown below.

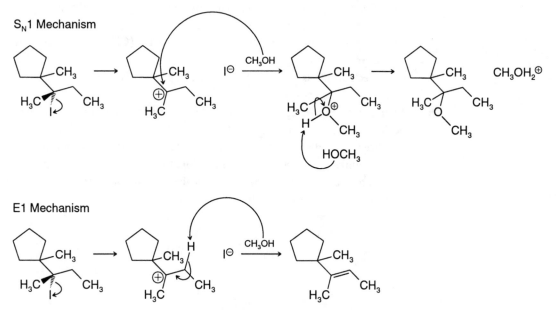

8.2

    a This looks like it needs to be an S$_N$2 reaction, based on stereochemistry. To ensure that, we choose a solvent that is aprotic, like DMSO, DMF, or acetone.

    b This needs to be an elimination reaction. The OH needs to be converted to a good leaving group, OH$_2^+$, using a strong acid like H$_3$PO$_4$. The solvent, H$_2$O, would then be protic. These conditions favor both S$_N$1 and E1, so heat is added to tip the balance in the favor of elimination.

    c This is a substitution reaction, where a Cl substitutes for an I. Therefore, the nucleophile must be Cl⁻. Because the product is a mixture of the R and S configurations at the stereocenter, it must be S$_N$1. To help promote S$_N$1, we should choose a protic solvent, like ethanol, CH$_3$CH$_2$OH.

    d This is an elimination reaction. Although the leaving group is good (I⁻), it is on a primary C atom, meaning that it would be difficult for it to undergo E1. Furthermore, we would like to avoid E1 if possible, in order to avoid possible side reactions of the carbocation intermediate. Therefore, let's help ensure that it goes by E2 by choosing a strong base that is a weak nucleophile, such as (CH$_3$)$_3$CO⁻. The solvent should be aprotic, like DMSO, DMF, or acetone.

8.3 Elimination requires that an H atom be bonded to the C adjacent to the C with the leaving group. There are no such H atoms in this molecule.

8.4 It must have been S_N1. The reason is that an S_N2 mechanism happens all in one step, and so the nucleophile MUST substitute at the C bonded to the L. Here it looks like the nucleophile attached at the adjacent C atom.

8.5 A mixture of S_N1 and E1 products. The nucleophile and base ($HOCH_2CH_3$) are both weak, favoring S_N1 and E1. The leaving group (Br) is good, which favors S_N1 and E1. The solvent is protic, favoring S_N1 and E1. The leaving group is on a tertiary C atom, which favors S_N1 and E1. The S_N1 organic product is $(CH_3)_3COCH_2CH_3$, and the E1 organic product is $(CH_3)_2C{=}CH_2$.

8.6 Same mechanisms as in Problem 8.5. The S_N1 product is $(CH_3)_3C-^{18}OH$, and the E1 product is the same as before.

8.7 When the leaving group leaves, what is left behind is a benzyl cation. If the charge is localized on that C atom adjacent to the ring, we would say that it is too unstable to have been generated, given that it appears to be a primary carbocation. However, there are three other resonance contributors that serve to have that positive charge shared on three C atoms of the ring. This gives substantial extra stability to that carbocation, helping to promote the leaving of L.

INDEX